CONTENTS

Contents

HIGH PEAK AIR CRASH SITES

Central Region

Pat Cunningham

COUNTRYSIDE BOOKS
NEWBURY BERKSHIRE

First published 2010

COUNTRYSIDE BOOKS
3 Catherine Road
Newbury, Berkshire

To view our complete range of books,
please visit us at
www.countrysidebooks.co.uk

ISBN 978 1 84674 219 4

Designed by Peter Davies, Nautilus Design
Produced through MRM Associates Ltd., Reading
Printed by Information Press, Oxford

'Get your eyes off that blessed SatMap and watch where you're going.'
Wies White, 2010

Other in-print books by the author:

Fiction

A Magnificent Diversion: series (Acclaimed by the First World War Aviation Historical Society's *Cross and Cockade International*)

Book One *The Infinite Reaches* 1915-16, ISBN 978-09556325-2-5
Book Two *Contact Patrol* 1916, ISBN 978-0-9556325-3-2
Book Three *Sold A Pup* 1917, ISBN 978-0-9556325-4-9
Book Four *The Great Disservice* 1918, ISBN 978-0-9556325-5-6

In Kinder's Mists (a Kinder ghost story) ISBN 978-0-9556325-0-1

'Though the Treason Pleases' (Irish Troubles) ISBN 978-0-9556325-6-3

Non-Fiction

Blind Faith: Joan Waste, Derby's Martyr ISBN 978-0-9556325-1-8

Contents

INTRODUCTION

The upland moors which feature in this book are among the most popular of Britain's walking areas. Each moor, consequently, wild though it may appear, is traversed by pathways which serve it so well that visiting one of the air-crash sites described in these pages rarely requires more than a 15-minute off-path diversion.

Even in good weather these moors have a mystique. But from the outset forget mystery. For not a shred of the unexplainable clings to any of the air tragedies recounted in this book. Sadly, in each case, the accident was due to human error, overwhelmingly to aircrew error; in just two cases, to errors made by non-flying supporting staff.

The essential cause of most crashes was that, flying at night and in cloud, the crews thought they were over a low-lying region and that a 2,000 foot reading on their altimeters meant, therefore, that they had that much clear air beneath them. What the altimeter was not designed to tell this hapless few was that their aircraft was already brushing a 2,000 feet above sea level, rock-strewn heather moor. And 'few', because the vast majority of the crews airborne on the same occasion did not stray over high ground, but landed safely. To the aviator-walker, therefore, imbued with a humble awareness of 'There but for the grace…', the true interest in such sites lies in the flight-safety lessons they taught others.

Some moors can be accessed from road level but even Kinder's uplands, though 1,300 feet above the valley, demand little more than an hour's climb. There are numerous starting points but such climbs invariably lead up picturesque water courses (cloughs), after which the walker is set for the day, the undulating rim path offering prospects of quite stunning grandeur.

Looking inwards, the heather-moors themselves have a less obvious charm. For a start, walking off-path through heather or bracken can be decidedly laborious, particularly as the vegetation hides boulders and peat crevasses. Then too, the flat areas tend to be marshy. Of even more note is that as the drainage develops so water channels (groughs) cut down through as much as 20 feet of peat to reach the bedrock below, leaving above them heathery islets (hags) and the impression of a vast, featureless wilderness.

Yet none of the moors is really that extensive, the centre of the Kinder plateau never being more than a 30-minute northerly or southerly walk from the unmistakable rim path. However, the moors do demand respect. Warm and still as it may be when kitting-up in the car park, the tops will be chillier and breezier, so sensible clothing is essential. Then again, when a mist falls it does so with startling rapidity, which means that a map and compass, and a working knowledge of how to use them, are equally essential. At the very least, if caught in mist, one should have confidence enough to set aside the inbuilt 'sense of direction' and trust to the compass needle!

Studying the proposed route on a map is bound to pay off. As is taking note of where the crash site lies in relation to the main path. And, of course, deciding upon the route beforehand lends itself to letting others know your proposed whereabouts: in an emergency the Mountain Rescue Service will thank your foresight. Certainly, do not expect a mobile phone signal, for moorland coverage is very poor.

In contrast, a GPS – global positioning device – especially one with a mapping display, is a great comfort on the featureless moors, even in clear weather, and most handsomely repays any expense when caught out in mist. But accept that mist causes the batteries to fail. Or so it seems. So take spares. Then again, with Google Earth you can walk the whole route before leaving home, and check out likely car parking into the bargain!

To visit the crash sites in this book many walkers will require only the guaranteed co-ordinates given, just the same, a walkers' guide follows each narrative. In contrast to the co-ordinates, the associated GPS-derived elevation is best regarded as merely a rough indicator. As for any timings given, then the youthful, lithe and long-limbed fell walker should be aware that they represent averages logged by a bulky, five-foot five-inches lone walker in his mid-seventies who will one day learn the wisdom of telling someone where he intends walking to. And then sticking to that route!

The air-crash debris

What the walker will see at most sites is a tiny pool of debris, often set in very broken terrain. For this reason, though the co-ordinates supplied are accurate, it may well be necessary to search about in order to locate the debris pool.

Debris, however, does not necessarily indicate the actual impact site. For when an aircraft bellied into the moor at cruising speed the wreckage was often sprayed over hundreds of yards; far further after an explosion. In other cases, when aircraft have struck rock faces, what debris remains tends to be pooled on lower ground. Then again, to avoid distracting later air searches, the salvage parties tumbled what wreckage they could not destroy into the nearest gully. Yet air-crash enthusiasts fight playground duels about a metre or two! That said, every site-seeking walker should be aware of the debt owed to researcher Mr Ron Collier (*Dark Peak Aircraft Wrecks*) who never enjoyed the benefit of GPS.

There is also much enthusiast wrangling about the wreckage itself**.** This walker-aviator, for his part, would like to see the Ministry of Defence (see Glossary) clear the moors of all such debris, conceivably, leaving a tasteful memorial marker to record the individual tragedy. Certainly, on the more accessible ground to the south and east of ancient Mercia – the area focussed upon by this series –, air-crash sites were returned to the plough the moment the salvage teams departed and only in rare cases was even an unofficial memorial raised.

It is reluctantly accepted, however, that many walkers have no interest in sites where nothing remains. Nevertheless, the fact is that, to the walker with a deeper aviation interest, some non-debris sites can be more worthy of a visit than those still-littered with wreckage. Certainly all sites in the series, whether fully or more slightly covered, have been equally proven by metal-detector findings, by contemporary photographs, and by verifiable witness reports.

Pat Cunningham

Area map showing the crash sites

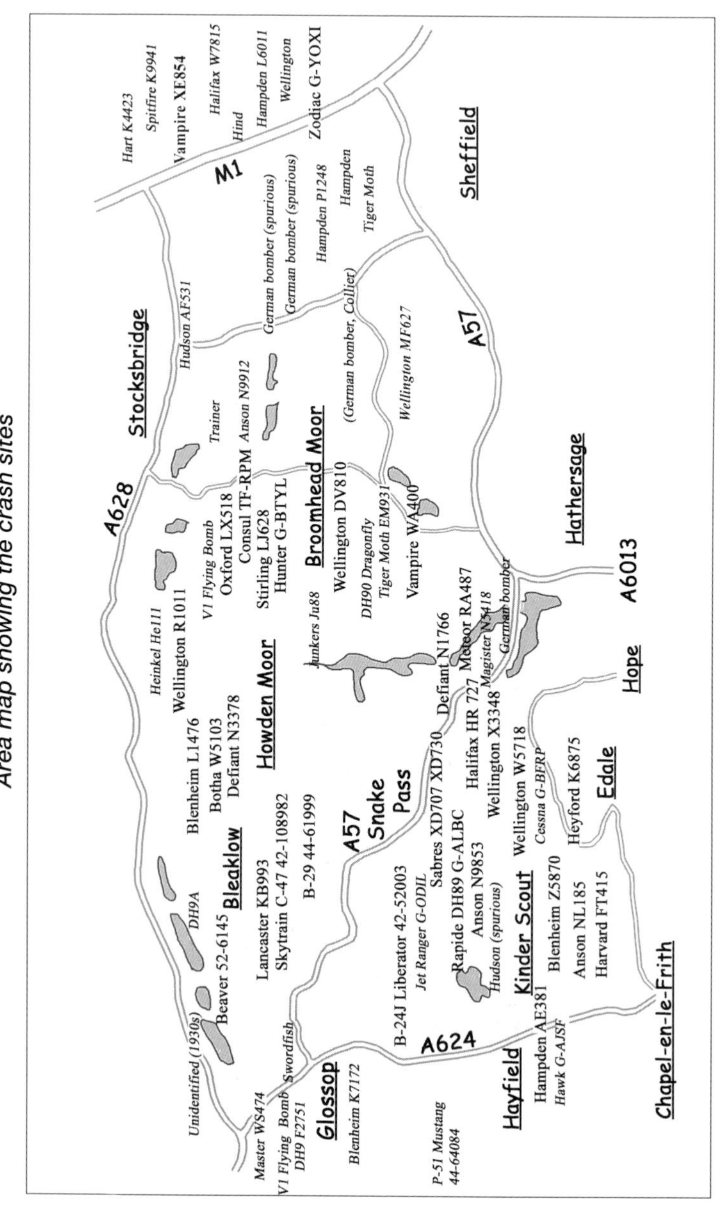

Sites with Debris

1

Consolidated Vultee B-24J Liberator 42-52003

Mill Hill, Little Hayfield

SK 05844 90610	500 m, impact site
SK 05731 90583	490 m, gully site
Unit and Station	United States Eighth Army Air Force, 27th Air Transport Group, 310th Ferrying Squadron, AAF582 (RAF Warton), operating from AAF590 (RAF Burtonwood)
Date	11 October 1944
Crew	Two, United States Army Air Force, both injured First Lieutenant Creighton R. Houpt, pilot Staff Sergeant Jerome M. Najvar, flight engineer

When an American operational squadron required a replacement aircraft it was often delivered by a ferry crew, based at aircraft storage and maintenance pools such as the Base Air Depot at Burtonwood, sited within easy reach of the Liverpool docks. Accordingly, when B-24J 42-52003 was to be delivered to AAF104 (American Air Force Station 104, the American designation for RAF Hardwick), just south of Norwich, the task was assigned to a pilot and a flight engineer who had not flown together before, indeed, who had never previously met. Not that this should have been any bar to a successful delivery, just so long as standard operating procedures were adhered to.

So it was that First Lieutenant Creighton Houpt, a ferry pilot with nearly 200 hours on B-24s, and Staff Sergeant Jerome Najvar, a flight engineer-cum-pilot's assistant, with a year's experience in ferrying B-24s and other bomber types, were arbitrarily brought together for the flight which was to result in the total destruction of the aircraft when it was flown into Mill Hill at an altitude of 1,640 feet above mean sea level.

Half cloud cover ('5/10') was forecast for the route with the main cloud base solid at 1,500 feet above the ground and a visibility of just 3,900 yards in haze. Then there was a gusty surface wind of 24 knots, itself a pretty fair indicator of considerably higher

Crashed 11 October 1944

The impact site

winds at altitude. Just the same, the flight clearance was that customarily issued for a ferry flight and required visual ground contact to be maintained throughout. The standard operating instructions for ferry crews were to go around areas of marginal weather, alternatively to put down at a suitable airfield, or to turn about altogether and return to base.

Having considered the conditions, First Lieutenant Houpt elected to fly at 2,800 feet, 'figuring', as he later stated in his report, 'that according to the map the highest point on course was 2,080 feet, which would leave plenty of margin to clear the hills.' He also recorded that before take-off he had set his altimeter to airfield elevation, or 76 feet. But the proposed 2,800 feet flight altitude – had he, indeed, maintained that height – would only have given him, at best, an 800 foot clearance, leaving little enough to compensate for the severe up- and down-draughts engendered by the strong wind.

On the other hand, First Lieutenant Houpt also declared that after taking off from Burtonwood he had set a south-easterly course; except that such a course would have taken him far south of Peakland's high ground, and nearly 20 miles clear of the 2,080 feet of Kinder Scout – and equally, of nearby Mill Hill! But all his declarations notwithstanding, what he had to admit to was that after 15 minutes' flying (representing about 29 miles), he saw hills; the crash site at Mill Hill being just that far, 29 miles, from Burtonwood, but directly *east* of it.

The middle gully

Leaving all that aside, just 15 minutes out from base First Lieutenant Houpt found himself 'flying on instruments in very rough air and little visibility'. So much for maintaining visual contact with the ground, as required by both his flight clearance and his standard operating procedures!

The flight engineer, acting as pilot's assistant, who is quoted as having been uncomfortable over the conduct of the flight – and, later, rather more than somewhat aggrieved – and who reportedly pointed out that the altimeter was only reading 1,500 feet and urgently advised a climb, has little to say in the official report. He notes that on returning from the standard 'wings and engines' check after take-off he had 'called the

pilot's attention to the 2,080 foot mountain on course,' and given an urgent thumb-up sign that he should climb. He concedes, nonetheless, that 'it seemed from the instruments that we had plenty of clearance'.

All they were afforded, however, was a momentary glimpse of a dark surface through a gap in the cloud, and although First Lieutenant Houpt immediately applied full power and pulled up, the aircraft struck the ground – to reiterate, at 1,640 feet above sea level, and with virtually that reading on his altimeter! – long before either action had time to take effect.

Although the aircraft was totally destroyed it did not catch fire and, as neither man had suffered incapacitating injuries, both were able to stumble clear of the wreckage. It must have been evident that they had crashed in a remote area, and that in the existing weather conditions any search would take some time to set in train. They decided, therefore, to make their own way down from the hills. As First Lieutenant Houpt reported, 'We walked downstream along a mountain creek until we hit a road.' Clearly they had descended the gully below the crash site and dropped into Hollingworth Clough, for eventually they reached the Hayfield-Glossop road, hitched a lift, and were taken to a public house where they phoned the tidings through to Burtonwood.

The Accident Committee could not fault either the aircraft or its instruments. They therefore found that First Lieutenant Houpt had been in error for 'holding the altitude of the ship at 2,800 feet on instruments, which would allow as safety margin less than 800 foot clearance of the ridges.' Further, they could see no justification for his having even attempted to continue the flight beneath the overcast.

Both crew members were briefly hospitalised with minor lacerations, although First Lieutenant Houpt additionally suffered a fractured jaw. His flight engineer's anger notwithstanding, who could doubt but that it was an injury occasioned by having impacted with something hard during the crash, rather than in the immediate aftermath!

Because of the relative inaccessibility of the crash site it was decided to set fire to the wreckage. Nevertheless, there are many stories of items being removed for years afterwards by locals; among other anecdotes being one recalled by Mr George Sherratt, of Glossop: 'My friend, Ken Bancroft,' he said, 'came from a tool-making family, but guns were always his passion. So on one visit he and another friend removed a machine gun. Along the way, however, they got tired of carrying it and hid it; except that, despite returning to search for it on several occasions, they could never find it again. But many years later, when Ken mentioned this to a keeper, it transpired that this same man had discovered its corroded remains, not that long before, and just where Ken had described secreting it.'

Such casual enterprise notwithstanding, even in 2010 a surprising amount of surface debris remained: debris almost continuously overflown by aircraft whose crews, unlike the hapless ferry duo, duly respect the Peaklands and fly at a safe height as they pass overhead in the process of letting down into Manchester's airport, just beyond the western skyline.

Visiting the Site

The most direct way to the site – one and a half miles (a 70-minute ascending walk) – is from the direction of the Grouse Inn on the A624 Hayfield-Glossop road, a lay-by just north of the inn offering adequate parking. Access to the moors begins opposite the junction with Monk's Road at SK 03250 90250, after which a well-established track offering fine moorland views leads over Burnt Hill towards Mill Hill. A paved section of the track leads to, and then skirts, the gully in which the site lies.

The other main route of something over two miles (taking of the order of 90 minutes) leads from the car park at Bowden Bridge, climbs William Clough – among the most inspiring of Kinder access routes! –, ascends Mill Hill, then descends on the same line to follow the paved route until it turns at a sharp angle. At this point an increasingly well-defined track leads directly ahead to the impact site. Moving south-west from the plateau leads into the debris-strewn gully which then re-connects with the paved track.

The lower gully

2

North American FMk.4 Sabres XD707 and XD730

Kinder Scout and Black Ashop Moor

SK 06926 89664	595 m, Kinder: initial impact point, start of debris trail
SK 07268 90236	470 m, Black Ashop Moor: wings, gear, with an engine in an adjacent grough
SK 07300 90100	480 m, Black Ashop Moor, two debris pools
SK 07548 90390	437 m, Black Ashop Moor, the second engine
Unit and Station	No. 66 Squadron, RAF Linton-on-Ouse, No. 12 Group, Fighter Command
Date	22 July 1954
Crew	Formating pilots, both killed Flying Officer James D. Horne, section leader (XD707) Flight Lieutenant Alan Green, formating pilot (XD730)

When Russian-built, swept-wing MiG-15 jet fighters were encountered in Korea in 1950, British manufacturers had nothing to match them. Home-grown transonic swept-wing fighters were under development to replace the straight-winged subsonic Meteors and Vampires, but until the Hunter arrived in 1955, Canadian-built Sabres filled the void. Most of the Sabres were based in Germany, but No. 66 Squadron was among those units equipped with them in Britain.

So it was that on 22 July 1954, four of the No. 66 Squadron Sabres were recovering to Linton-on-Ouse, near York, after a high-level interception sortie flown in the course of a major annual evaluation war-game. For the descent, the formation leader had split his section into pairs, each pair entering cloud independently at 12,000 feet. Some time later, as his pair passed 5,000 feet, still in cloud, the overall leader transmitted an advisory warning to Flying Officer James Horne, now leading the second pair, against descending below 3,000 feet on their present heading: the more realistic safety height of the future was to be 3,800 feet!

The main terminal site on Black Ashop Moor and memorial, looking along the line of flight

The foursome had noted already that Flying Officer Horne's radio was weak at times, so he may not have heard the warning; certainly, he did not acknowledge. Nothing more was heard of him, or of his number two, until three days later when a walker came upon a body on The Edge, high above Black Ashop Moor.

Until this discovery, bad weather had hampered the search, although the keeper of the Kinder Reservoir had reported being alarmed by two jet fighters roaring at very low level towards cloud-covered Kinder. Despite his concern, however, he had heard no subsequent impact.

These aircraft had undoubtedly been the two Sabres, although what made Flying Officer Horne take his number two that low will never be known. A likely scenario, however, is that he saw a clearance below him and dropped into what turned out to be a 'sucker's gap' – a beckoning clearance which then closed in around him. Certainly, in his evidently hasty pull for a safe height, he managed to clear the edge of Kinder, but equally evidently something untoward happened after he had done so, for both aircraft struck the ground in a single impact point not many yards into the plateau.

To conjecture further, although Flight Lieutenant Alan Green, the number two, was more experienced than his leader, he was still settling in having been posted from another squadron. It could be then, that, caught out by the hastily initiated transition from level flight to very steep climb, he had – understandably – twitched just that little bit, causing his wingtip to lock with his leader's tail. Or what is equally likely, bearing in mind that both aircraft were in very steep climbing attitudes, is that in reaction to his leader's over-hasty pull Flight Lieutenant Green had pulled even harder, got high

Crashed 22 July 1954

and, being momentarily unsighted from his leader, had collided when blindly pushing back into position.

The court of inquiry, however, did not treat with such speculation, finding only that Flying Officer Horne, as section leader, had failed to observe the area safety height – which he should have been familiar with regardless of any missed transmission – although it found some little mitigation in his faulty radio. It specifically noted that no blame was to be attached to Flight Lieutenant Green, whose sole responsibility had been to formate upon his leader.

The walker who had stumbled upon the tragedy was Mr Graham Atkin. 'It was the Sunday of that week, at about midday,' he recounted, 'and I was heading for Fairbrook Naze. I'd left Ashop Head some way back when I saw something white ahead of me. At first I thought it was snow – but in July? It was, of course, an only partly-deployed parachute. And then I saw the body of the pilot. He'd clearly been there for some time, and equally clearly, although this was the first dead person I'd ever seen, there was nothing to be done for him. So I angled my way down into Black Ashop Moor, heading now for the Snake Inn and initially stepping over scores of cannon shells and bits of metal. I didn't see any of the major wreckage down in the valley, so by angling off I'd obviously passed to the right of that. Once at the inn I dialled 999 and reported the crash. Then I caught the bus into Glossop [the good old fifties!], to the police station. They'd made phone calls, among them to the RAF Mountain Rescue Unit at Harpur Hill, and now they took me to Hayfield where I led them back up via William Clough to where the pilot was. A policeman took down my statement in his notebook. Then, I suppose, I left them all to it, and went on down again: how fit I must have been back then!'

He paused. 'Some days later I had to attend the coroner's inquest at Marple. And really, that was that. I've always supposed the two planes simply crashed into the hill in the cloud, but, as it happens, only a couple of years ago I was leafing through a book on Derbyshire where the author had it that they had been led into the cliff by a ghost aeroplane! What a lot of rubbish! But then people tend to make up these stories of ghost aeroplanes in the Peak.'

'I remember,' he added, 'in about 2000, Alan Jones visiting to talk me through what I saw that day. Apparently he'd actually found the pilot's ejection seat.'

In fact, researcher Mr Jones had spent a considerable time examining the variously scattered sites of the two Sabres. Indeed,

One of several debris pools on Kinder

The cockpit of one of the Sabres, recovered from the moors by researcher Alan Jones

in October 2001 he would break an ankle in the process and have to be stretchered off Black Ashop Moor! 'We found that particular pilot's ejection seat,' he said, 'just off the plateau, jammed into the first tiny gully on the Black Ashop Moor side of the path. Although by that time floods had packed it full of rocks.'

Given the juxtaposition of body and ejection seat it is possible that the pilot had tried to eject but that, early ejector seats not having the ground-level, zero-speed capability they later acquired, the attempt had been made in descending flight and fractionally too late. It seems more probable, however, that he had been unconscious throughout and that he and his ejection seat had been separated as the two aircraft continued to disintegrate; a more kindly sequence of events, and hopefully the correct one.

Much evidence of this accident still remained in 2010, the debris trail stretching from the impact point, just inside the edge of the Kinder plateau, for a thousand yards north-eastwards into the depths of Black Ashop Moor, where major components lay.

Visiting the Site

The initial impact area is encountered some five minutes (400 yards) after scaling the Ashop Head Scarp with Sabre debris to the left-hand edge of the popular Kinder Rim Path. An unchallenging ten-minute detour (under 400 yards, but over broken ground) towards the northern rim path will reveal several more debris pools. Passing a little beyond that rim path to see over the convex slope affords a bird's-eye view of the various terminal sites in Black Ashop Moor, 500 feet below.

The excursion also gives a panoramic overview of the moors towards Doctor's Gate and Bleaklow which well repays the effort involved, particularly for those not yet familiar with the dramatic nature of Kinder's less frequented northern edge.

Regarding the terminal sites on Black Ashop Moor, these are best approached from the Snake Path. Ideally, from the region where the infant River Ashop is joined by the flow from Within Clough, for from thereabouts faint footpaths cross the rough and marshy moorland to reach the main site. For the truly intrepid, climbing back along the line of flight and up onto Kinder would take half an hour of pathless ascent.

5

Ringway (Manchester)
Last-directed heading maintained
Cluther Rocks

Handley Page Hampden Mk.1 AE381

Cluther Rocks, Kinder Scout

SK 07793 87485	614 m
Unit and Station	No. 50 Squadron, RAF Skellingthorpe, No. 5 Group, Bomber Command
Date	21 January 1942
Crew	Four, all killed Sergeant Royal George Heron, Royal Australian Air Force (RAAF), pilot Sergeant Walter Chantler Williams, RAAF, navigator Sergeant William Thomas Tromans, RAF, wireless operator Sergeant Sidney Albert Peters, RAF, wireless operator/air gunner

At its manufacturer's declared maximum speed of 254 mph (221 knots), the Hampden was the fastest bomber of its time, but once war had broken out it proved too vulnerable in the face of enemy fighters and was quickly switched to night bombing. It did serve, just the same, as the platform for the first two Bomber Command VCs in late 1940. For Sergeant Pilot Royal Heron and his crew, however, airborne on the night of 21 January 1942, there was to be no such kudos.

Having taken off on a night cross-country exercise from RAF Skellingthorpe, near Lincoln (a satellite airfield of RAF Swinderby), they ran into blizzard conditions, became lost, and crashed into Cluther Rocks on the western rim of Kinder Scout, 50 miles off their planned course. There were no survivors.

Sergeant Heron had been trained in Canada under the Empire Air Training Scheme, and would hardly have been fully acclimatised to United Kingdom conditions; nor was his navigator, Sergeant Walter Williams, a fellow Australian, all that experienced. Offsetting this, the court of inquiry was assured that before take-off the crew had been fully briefed on radio beacons and also in obtaining such emergency in-flight navigational assistance as 'Darky' (see Glossary); additionally the use of occulting –

essentially, direction-indicating – searchlights had been covered. In fact, though low in experience, on accepting that they were lost Sergeant Heron had sensibly had Sergeant William Tromans, his wireless operator, contact the direction-finding (D/F) facility at Ringway (subsequently Manchester International Airport) for a series of bearings, obtained and relayed by morse code, which enabled them to be homed towards that airfield.

A popular account (Collier/Wilkinson, 1979, *Dark Peak Aircraft Wrecks 1*, p.106) has former airman Mr Herbert Ward making reference to a radio beam. And it is possible that, with the products of both the Avro and Fairey factories to be flown off locally, to say nothing of its increasingly important paratroop commitment, Ringway might well have received a Standard Beam Approach (SBA) installation (see Glossary) that early on. Yet even had the facility existed, it hardly seems likely that this particular crew would have had the

The impact site, above the rim path, looking towards Kinder Reservoir

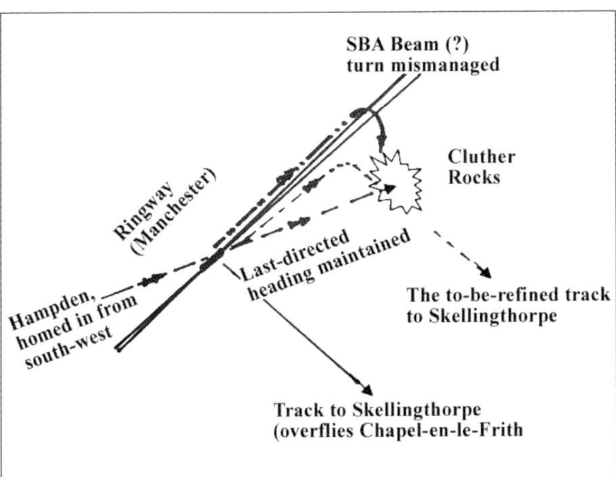

The track flown by Hampden AE381

training necessary to utilise it. One suspects, rather, that the former airman – by his own admission only getting the gist of things from wireless-operator colleagues – introduced the knowledgeable-sounding term 'beam' to add verisimilitude to his account, and that this crew, with 'neither pilot nor navigatorveryexperienced', as the RAF investigators observed, never so much as attempted to access a radio beam.

The beam issue aside, the homing Hampden was eventually heard passing overhead Ringway. At this point Sergeant Heron would have received the customary advisory of the period, 'Engines overhead'; a call which would have immediately relieved his crew of their uncertainties regarding their location, while at the same time requiring them to assimilate the fact that they were a full 50 miles from their intended track.

It is not known what decision Sergeant Heron made at this juncture. He could well have decided that enough was enough, and elected to land at Ringway. But that would have meant embarking upon a cloud-break procedure, or alternatively, given that SBA was indeed available, upon a full-blown beam approach, neither option an easy undertaking in the blizzard conditions obtaining.

Then again, giving up and landing at Ringway would have meant his crew facing the bantering derision of their peers on their eventual return to base. Best, surely, for Sergeant Williams, the navigator, to simply plot a revised course for Skellingthorpe, where the weather could hardly be worse, and where they would arrive with no suspicion cast that they had ever got themselves quite so totally lost, if a little late.

The navigational problem facing them was straightforward enough, for as Sergeant Heron battled the gusts, Sergeant Williams would have ruled off a track from overhead Ringway to Skellingthorpe of some 100° True – eastish – at a distance of 43 nautical miles. Given a workaday cruising speed for the Hampden, and adding on the healthy south-westerly (tail) wind ahead of the blizzard, it was something less than 20 minutes' flying time.

Whatever on-board decision was made, however, deliberations must have taken an appreciable time, for after a while the tower, mindful of the high ground to the north and east of Ringway and aware that the aircraft was flying at only 2,000 feet, directed it to turn about and return to the overhead; except that nothing more was heard from the Hampden.

Discounting the possibility that this had indeed been a beam-trained crew who fatally mishandled the inbound procedural turn – no anachronism there, for both procedure and terminology date back to 1938! – then two other scenarios present themselves. The first is that, having been homed to the overhead, the Hampden crew merely maintained the last directed heading; which had to be something rather more easterly than north-east. The second, and the kindest, is that they had reassessed their position and had already turned onto a roughly estimated easterly heading for Skellingthorpe, intending to refine this as they neared their base. Except that the final result of all three possibilities was the same, for with just 2,000 feet showing on their altimeter – undoubtedly then regarded as an adequate height for their original route and which, on track for Skellingthorpe from actually overhead Ringway, would still have been safe enough! – the Kinder massif, at 2,088 feet, and Cluther Rocks, only slightly lower, lay directly in their path.

The crew died in the intense fire caused on impact, as rescuers who had struggled through the blizzard – including the impressed Airman Herbert Ward – were dismayed to find. But another discovery, reported by the farmer from Hill House Farm (just east of the Bowden Bridge car park), was a widespread scatter of propaganda leaflets, which understandably gave rise to the belief that the aircraft had just returned from a leaflet-dropping operation. Only, air-dropped leaflets have a tendency to not only plaster themselves over the aircraft employed, but to secrete themselves in and about it, and these proved to be the leftovers from some previous sortie.

Visiting the Site

The crash site lies on the Pennine Way path between Kinder Downfall and Kinder Low. On passing Cluther Rocks from the direction of the Downfall the path changes levels for the short ascent to the Kinder Low plateau. Just before it does so, and only feet above and to the left of the path, there are two debris pools, the rocks themselves being liberally peppered with tiny fragments of wreckage. As an added incentive to visiting the site, scrambling to the top of the rocks enhances somewhat the already spectacular view from this rim path.

In 2005 a vandalised wooden *in memoriam* cross was replaced by a slate slab which, regrettably, makes no attempt to adhere to the established hierarchy in listing the crew. Back in 1991, however, in more pleasing sympathy with the site, aviation enthusiast Mr John Fairbrother, from Stalybridge, was 'just poking about among the rocks with a stick', as he recounted, when he came upon the remains of a watch engraved '*R.G. Heron, RAAF*'. After much perseverance by aviation author Mr David W. Earl, this was eventually returned to the pilot's family in Australia, undoubtedly furnishing some tangible closure to the loss they had suffered all those years before.

4

Avro Anson Mk.9 NL185

The Cloughs (north-west of Upper Booth), Edale

SK 08903 86696	506 m, impact site
SK 08873 86570	465 m, engine site
Unit and Station	Headquarters Bomber Command Communications Flight, RAF Halton, Aylesbury
Date	23 November 1945
Crew	Pilot, killed Wing Commander Richard Douglas Speare, DSO, DFC and Bar, Croix de Guerre with Palm

Avro Anson NL185 was on the charge of the Communications Flight at RAF Halton as a station runabout when it crashed into The Cloughs, on Kinder Scout, on 23 November 1945, at a height of 1,700 feet above sea level. The pilot, Wing Commander Richard Speare, was no stranger to the Anson having been a Service pilot for nine years. Nor had those years been uneventful, for he had enjoyed 'a good war' on bombers, culminating in operating No. 138 Squadron's special-duties Halifaxes from RAF Tempsford, delivering agents and supplies to the Continent for the Special Operations

The pilot, Wing Commander Richard Douglas Speare, DSO, DFC and Bar, Croix de Guerre with Palm **(courtesy of Chris Steele)**

The debris pool at the impact site

Executive (see Glossary). Further, his outstanding operational service had gained him a Distinguished Service Order, a Distinguished Flying Cross (twice awarded), and a Croix de Guerre with Palm; all these following on from a well-earned Mention in Despatches.

Since training in 1936, Wing Commander Speare had flown some 2,800 hours, considerably more than the average Service pilot of that era. On this particular occasion, however, stowing aboard all his kit for an anticipated leave, he had submitted a flight plan to ferry a machine from Halton to Feltwell (near Ely), a distance of 70 miles on a north-easterly track (050° True). Why the aircraft flew a north-*westerly* track (330° True) and crashed some 118 miles *north-west* of Halton, remains something of a mystery. Indeed, it was a lifetime source of disbelief to some who knew Wing Commander Speare's capabilities. But then a crash site, once interpreted, tells it as it is.

The court of inquiry found that the aircraft had indeed been vastly off course when it flew into cloud-covered high ground. The members did suggest that Wing Commander Speare might possibly have set the wrong compass course, but the RAF accident report summarising the investigation does not record whether investigators at the scene actually found the compass-setting ring locked to some north-westerly, rather than some north-easterly, heading. Just the same, mis-setting the required heading in this way was one of several only too common navigational compass errors.

A similar error was to misread the pre-calculated heading when setting the compass. So common were such errors that the 1941 RAF *Air Navigation* manual devoted an admonitory section to 'Setting Reverse or Incorrect Courses', directing that 'The Navigator must keep a wary eye on…the Pilot, who, for various reasons, may not be steering the requisite Course', and further, 'It is advisable to look repeatedly over the

Pilot's shoulder at the Course being steered…'. Poor solo ferrying pilot then, with no one to look over his shoulder!

Aside from the directional error, however, there was poor visibility between Halton and Feltwell that day, a meteorological condition which suggests only light winds. It was also recorded that the actual and the flight-plan winds were virtually identical. Which meant that the 70-mile flight to Feltwell would have taken an Anson only 22 minutes. Yet even when Wing Commander Speare made the decision to descend blind he must have been flying for a good 14 minutes longer than that.

It so happened that this particular machine had been stripped of its radio equipment, so that besides a compass Wing Commander Speare had only his map and mental dead-reckoning to assist him; little enough in cloudy, low-visibility conditions. Further, having spent his nine years of aviating on bombers, he would have become reliant upon the skills of a dedicated observer (navigator) and would not have been at all well-versed in the demands of single-pilot operation.

The court of inquiry dutifully proffered the lack of radio equipment as a contributory factor. But whether Wing Commander Speare would have called for radio assistance had a radio been fitted seems problematical, it being far more likely that he never doubted himself to be anywhere but in the vicinity of Feltwell. Or not until the last

Researcher Sean Moran at the engine site

instant, for it was clear from the disposition of the wreckage that he had seen the ground and desperately tried to avoid it; 'but was too late', as the crash report has it.

In this instance then, the often fraught decision to let down through cloud must have seemed unexceptionable to Wing Commander Speare, for the terrain in the Feltwell area was nowhere much more than 30 feet above sea level, even had he – with his altimeter still registering against the 370 feet above sea level datum set at Halton – flown on past his estimated time of arrival. So while the unyielding slopes that constitute the Derbyshire crash site send a shiver down the spine of anyone visualising those final cloud-shrouded moments, the last thing on earth – a suddenly sobering phrase – Wing Commander Speare would have expected was to fly into a hillside when his altimeter was still reading some 1,700 feet. Only in truth he was now a full 95 miles north-west of those safe, flat, sea-level fens around Feltwell.

So it was that his body, made singular by his multiple rank-braids and the colourful array of medal ribbons, was duly found beside the aircraft high on The Cloughs amid the boulder-strewn grasses, a scatter of wreckage, the tumbled contents of his personal suitcases, and his Service cap.

Visiting the Site

Initial access to the area is most popularly obtained by ascending to Kinder either by the Jacob's Ladder path or by the Grindsbrook or Crowden Clough routes. On the other hand those with strong hearts and wiry sinews might care to bare-facedly breast the slope, ascending the unnamed water channel (at SK 08789 86344) just upstream from the Jacob's Ladder Bridge.

Both the sites associated with the Anson crash are actually on that unnamed watercourse, high on the steep, grassy slopes directly below the Wool Packs and immediately above the foot of stone-stepped Jacob's Ladder.

The sites themselves can be approached from above – with due deference to the gradient – by descending for some 300 yards anywhere between the features of Pym Chair and the western end of the Wool Packs rock formation. On the descent the sweeping views ahead, from Brown Knoll to Lord's Seat and beyond are magnificent, but best taken in before leaving the rim: once committed, the ground actually underfoot has to be the priority! Again, during the descent, the reed-fringed debris pool of Harvard FT415 (see below) may well be seen. The Anson sites, however, will only be discovered at the last moment. An alternative descent is by the yet-faint foot trail following the previously mentioned minor water channel to the eastern side of Wool Packs. (In fact, it leaves the rim path at SK 08977 87011 618m.)

In early 2010 a fair pool of collected Anson debris remained in a hollow beside the water course. One of the aircraft's two Armstrong Siddeley Cheetah engines, idly rolled over by passing walkers, had long begun a gradual descent. For a good ten years it had rested in a bed of rushes some 450 feet downstream but in May 2010 it was located 45 feet lower still and virtually hidden in a deep peat hole.

5

North American Harvard Mk.2B FT415

The Cloughs/Wool Packs, Edale

SK 08888 86897	580 m, impact point
SK 08895 86835	560 m, debris pool
Unit and Station	No. 22 Flying Training School, RAF Syerston, No. 23 Group, Flying Training Command
Date	14 January 1952
Crew	Pilot, killed Midshipman Brian Farley, Royal Naval Air Service, pupil pilot

Of the 17,000 Harvards built during the course of the Second World War the Royal Navy acquired 236, receiving its first in 1943. Even before that, however, Naval pilots had been trained on the type in various Dominion countries under the Empire Air Training Scheme. Indeed, the Harvard was still the prime advanced trainer in 1952, when pupil pilot Midshipman Brian Farley was on his wings course at RAF Syerston, near Newark, in Nottinghamshire.

On 14 January 1952 Midshipman Farley was briefed to fly a multi-legged navigational cross-country exercise from RAF Syerston which would include landing away from base at RAF Kemble, near Cirencester, Gloucestershire. But Midshipman Farley's machine did not arrive at Kemble and nothing was found of it by an air search which widely bracketed the prescribed route.

It was not until six days later that gamekeeper Arthur Lowe discovered its wreckage high on The Cloughs, in Derbyshire's Vale of Edale, a location so far off the track set for the navigational exercise that the search had failed to cover it. The aircraft had crashed and burnt on impact, killing Midshipman Farley. In later years, as farmer Mr Roy Cooper of Highfield Farm, Upper Booth would verify, 'Arthur would talk of finding the pilot still strapped into the cockpit. "He was totally mummified," he would say, "more like a dummy than a man." '

Mr Cooper had himself, in fact, seen the Harvard 'flying very low indeed, up the

Inset: The impact site Main image: The debris pool

railway line towards the tunnel', but had not been called upon by the investigators. Another witness, however, told the inquiry that as the Harvard had approached the Cowburn Tunnel it had abruptly pulled up into a steep climb. This witness's impression had been that the pilot had not known that the railway entered a tunnel and that on realising it he had been forced to start a climbing turn. The tragedy being that anything other than a complete reversal of direction faced Midshipman Farley with ridges demanding a 1,500 foot zoom to clear their summits. Clearly unaware of this, the luckless student pilot opted for a moderate right turn, then ran out of space and rate-of-climb alike to impact just below the ridge.

Crashed 14 January 1952

The members of the inquiry discovered that Midshipman Farley, with just 41 hours of solo experience, had not submitted his flight plan to an instructor before flight. Further, in view of the fact that he was so far off track and heading for Manchester, they reasoned that rather than adhering to his authorised route he had decided to overfly Stockport, his home town. Not the first pilot to decide to 'beat up the folks', and very far from the last. But most transgressors do not kill themselves, or destroy their aircraft.

Visiting the Site

The rim path serving both Pym Chair and Wool Packs can be reached in a number of ways: from the Pennine Way up stone-paved Jacob's Ladder, from an ascent up Grindsbrook or Crowden Cloughs, or, for the stout-hearted, by climbing directly upslope, and essentially off track, following the unnamed water course issuing at SK 08789 86344, just above the Jacob's Ladder Bridge.

The Harvard's impact point, a singular slab of slanting rock 150 yards below the rim, unmarked but proven by contemporary photographs, is just below Pym Chair and directly above the sparse pieces of Harvard debris resting in a reed bed some 70 yards – 50 feet vertically – below. Due care should be taken when descending from the rim path for the going is over rough, trackless hummocky grass on very steep ground. Adjacent to the debris some faint lettering chiselled on a rock by the salvage team was still discernible in early 2010. Little enough to show for all the exuberance and high ambition so suddenly extinguished on that brisk January day.

The alternatives for the return depend upon the route in hand. Back up to the rim path, for instance. But descending by the unnamed water course is a feasible, if knee-punishing, way of reaching the Pennine Way, and Edale; the skyline view of the crags of Wool Packs which this descent affords makes the effort well worthwhile. (Offered merely as a sop to those misguided enough to take this descent: the craggy skyline can be appreciated far more comfortably from the foot of Jacob's Ladder, on the easier-on-the-knees Pennine Way!)

The Harvard site, of course, would undoubtedly be taken in while visiting the twin sites of Anson NL185 (see above) which crashed only yards below the Harvard.

6

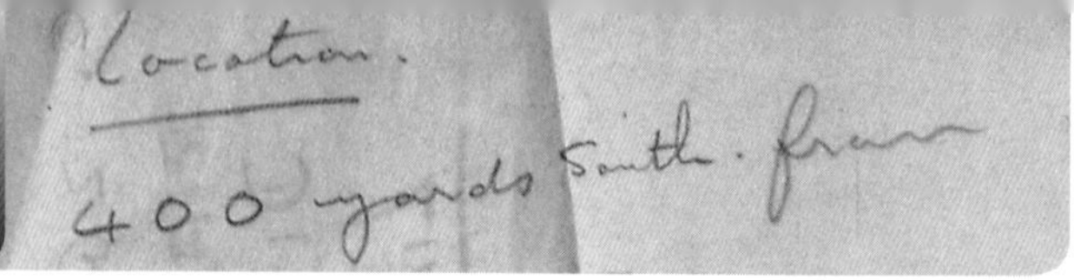

Bristol Blenheim Mk.4 Z5870

Below Crowden Tower, Edale Moor

SK 09437 86747	552 m
Unit and Station	No. 12 Group Anti-Aircraft Cooperation Flight, RAF Digby, Fighter Command
Date	3 July 1941
Crew	Two, and two passengers, all killed: Sergeant Nicodem Płotek, Polish Air Force under British Command, pilot Aircraftman Second Class Wilfred Cottom, RAF, wireless operator/air gunner Aircraftman Ron Place, RAF, passenger Aircraftman William Franklin Kidd, Royal Canadian Air Force, passenger

One of the roles undertaken by the Blenheim after being withdrawn from service as a first-line bomber was that of providing an exercise target for the guns, searchlights and radar of Anti-Aircraft Command. On 3 July 1941 when it crashed, however, Blenheim Mk.4 Z5870 was merely being ferried to the maintenance unit at RAF Ringway (Manchester) for routine servicing. In view of this its pilot, Sergeant Pilot Nicodem Płotek, of the Polish Air Force under British Command, agreed to take along two airmen who had begged a ride.

The Polish Archives show that before the September 1939 Campaign – with the Germans invading on the first of that month and the Russians on the seventeenth –, the twenty-four year old had logged nearly 400 hours while serving with the Third Polish Air Regiment. Just prior to the onslaught, however, he had transferred to the Bomber Brigade where he flew operations in antiquated three- and four-crewed machines. In 1941, sobered by that experience, his wish was that the RAF would reassign him to fighters. 'Then,' he told the Tyas family, with whom he was billeted, and who became his nominated next of kin, 'I can kill Germans and if I have to die it will be alone, and no longer responsible for other people.'

In fact, on arriving in Britain on 9 February 1940, Sergeant Płotek had been accepted into the RAF Volunteer Reserve (RAFVR) at Eastchurch, on the Isle of

Sheppey, and given the rank of aircraftman-two (AC2), remustering next day from general-aircraft hand to U/T (under training) pilot. This reflects the stance the British took with Polish airmen coming to Britain: all commissioned Poles were to become pilot officers, those from warrant officer downwards becoming AC2s; further, no credit was given for any operational or other service before arriving in Britain. These rules were based upon the fear that low Polish morale – presumed after their defeat in Poland – might affect British fighter units, and that the language barrier would prove insurmountable in fighter operations. Only when the Polish fighter pilots in France had shown their mettle did the British realise that they had 70 Polish pilots who could aid in the Battle of Britain.

AC2 Płotek, as a bomber pilot with recent experience, had duly undergone acclimatisation training, being introduced to RAF techniques while variously passing through No. 3 (Polish) Wing at Blackpool, No. 15 Elementary Training School at Dumfries, and No. 18 Operational Training Unit at Hucknall, Nottingham.

Appointed acting sergeant (paid) he had then been posted to Headquarters No. 110 Anti-Aircraft Cooperation Unit (AACU) Wing at Manchester's Ringway before being assigned initially to No. 6 AACU at Kirton in Lindsey and then, on 29 May 1941, to No. 12 AACU at RAF Digby. On 6 August 1940, with the re-formation of the Polish Air Force as the 'Polish Air Force under British Command' he had been formally discharged from the RAFVR.

On 3 July 1941 Blenheim Z5870 was to be flown from RAF Digby (north-east of Cranwell) to RAF Ringway, an 80-mile flight which would have taken under 30 minutes. As it was, the aircraft crashed on The Pike, below Crowden Tower, Edale, some 4 miles to the right of its straight-line track.

The machine was discovered by Gunner John Hamer of the Royal Artillery, who spotted it while out walking; escaping, as was his off-duty wont, from his searchlight post beside the mouth of the Cowburn Tunnel. As he was to tell researcher Mr Alan Jones, in a letter dated January 2000, he had ascended the track of Chapel Gate from Barber Booth and was making his way around the head of the valley via Colborne and Brown Knoll, when he saw the aircraft high on the distant moor. So it was that instead of descending to Lee Farm by the Jacob's Ladder track, as had been his intention, he had continued around Kinder's northern rim towards Noe Stool and in due time had come upon the Blenheim. He remembered that although it had gouged a grassy north-westerly furrow into the heavily fissured peat of the plateau, it had seemed virtually intact. Just the same, all four of its occupants were dead.

The official investigators were unable to determine a definite cause for the crash, and although engine failure was considered it was eventually discounted as unlikely. It was held, therefore, that the pilot had encountered lowering cloud and had descended with the cloud base in a vain attempt to maintain visual contact with the ground.

Because the aircraft had largely retained its basic integrity little wreckage was shed. Not surprisingly, therefore, the location was lost for many years, with enthusiasts unrewardedly searching both the moor above Crowden Tower crag and the grassy slope below it – 'moor' being the term local farmers habitually reserve for the upland plateau.

In 2003 the earliest locational evidence seemed to be the testimony of Police Constable J.A. Coleman, of Hope, who told the coroner that the Blenheim was found 'approximately 300 yards west of Crowden Tower, Edale'. Yet such a position would have been out of Gunner Hamer's line of sight as he traversed the Colborne-Brown Knoll ridge. There was a second on-the-moors position, but this too proved to be unsighted. The location supplied by Mr Ron Collier (Collier/Wilkinson, 1979, *Dark Peak Aircraft Wrecks 1*, p.153) of '092 871' – that is, SK 09200 87100 – seems to have been taken from the policeman's report, for it too could not have been seen from the Brown Knoll heights. Then again, Collier/Wilkinson, while supplying co-ordinates which put the site on the summit moor, introduce their narrative with: 'crashed below Crowden Tower'.

Back in 1941, though, the RAF summary crash report had itself confused the issue, with the original location of 'Crowden Tower' being crossed through in favour of the handwritten amendment, 'Lee Farm'. (As it happens, this summary also records that the aircraft hit a hill at 2,800 feet, whereas Kinder Scout lifts to only 2,088 feet!)

On the other hand, Gunner Sydney Becket's evidence to the Coroner's Court accorded well with the constable's statement that the aircraft lay 'approximately two and a half miles from where the road ends at the Lee Farm, Edale'; for routing from Lee Farm and the Jacob's Ladder path to pass Noe Stool on the northern rim gives just that distance. Then again, when approaching from that direction Crowden Tower is on a level with the moor and its crag cannot be seen. It seems certain then, that their point of reference was not the Tower, but the interposing 619 m outlier of the Wool Packs, so accounting for the amendment to the RAF crash report.

Before the previously mentioned letter of January 2000, researcher Mr Alan Jones had actually visited Mr Hamer who had then described how he and the other young soldiers detailed for sentry duty reached the site: 'We would go by lorry to Upper Booth and then continue on up a track until we came to a place where a high bank [hillside] was on the right… Up this we scrambled … I have the impression it was not too far to the crash site, say 400 to 500 yards ... The ground was fairly level and it must have been grassy for I have no feeling of heather.'

It then transpired that early in the eighties part-time ranger Mr Ron Weeks had told researcher Mr John Ownsworth that the aircraft crashed 'on The Pike' – a location which retired Area Ranger Gordon Miller confirmed was the local name for the spur leading down from Crowden Tower to Lee Farm. Mr Weeks also held that what debris there was had been removed by enthusiasts between 1970 and 1971.

In the absence of debris, any more positive determination of the crash site had to be speculative. In 2007, however, the matter was finally cleared up when Mrs Barbara Morris (née Tyas) – the daughter of Sergeant Płotek's 'adopted' family – produced

Crashed 3 July 1941

The plateau below Crowden Tower, the impact site

a slip of paper handed to her father (as nominated next of kin) by the Sergeant's commanding officer just three days after the event. This specified: *Location, 400 yards South from Crowden Tower, N of Lee Farm, Edale.* With the Tower being at SK 09435 87112, this gives the terminal location as SK 09437 86747.

For the record, in November 2007, reacting to this contemporary information, a three-party metal-detector search still failed to turn up any physical evidence.

Inset opposite: Details of the crash's location, as supplied by Sergeant Płotek's commanding officer three days after the tragedy

Crowden Tower seen from the impact site

Crashed 3 July 1941

Visiting the Site

There is no debris to see at this site but it more than qualifies for inclusion in this something-to-see section because of the number of red herrings which had to be sniffed out and the sheer foot-pounds of energy – to say nothing of the time – expended in determining its whereabouts. At least the climb to the site offers two very worthy alternative routes to Kinder Scout, both conveniently emerging near other crash sites.

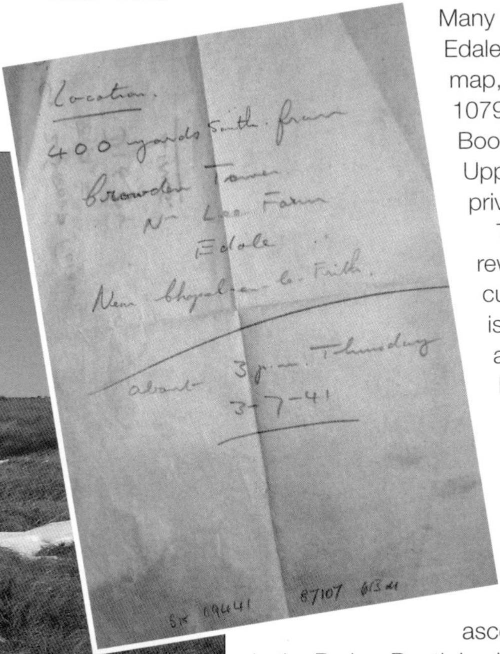
Location.
400 yards South from
Crowden Tower
Nr Lee Farm
Edale
Near Chapel-en-le-Frith
about 3 p.m. Thursday
3-7-41
SK 09441 87107

Many walkers will start from the car park at Edale. Ample parking, however, marked on the map, is also available in the large lay-by at SK 10797 84713, just along the road from Barber Booth. In fact, the road continues beyond Upper Booth but from there on it is gated to private vehicles.

The easiest route, and the most scenically rewarding, is that taken by the rescue-cum-salvage parties who followed what is now the Pennine Way past Lee House and Jacob's Ladder. Just short of Kinder Low this branches right for the Noe Stone and the Wool Packs before reaching the area of the site, some way short of Crowden Clough. Those following this classic walking route will appreciate that Crowden Tower (SK 09435 87112) never becomes a feature!

A rather more adventurous route ascends Crowden Clough. For those parking in the Barber Booth lay-by, this heads north immediately on leaving Upper Booth. Walkers starting from Edale, however, would branch right from the Pennine Way having skirted Broadlee-Bank Tor. The Crowden Clough ascent is, in fact, one of the finest routes up Kinder and one, moreover, which shows the Tower to good advantage throughout. Like so many cloughs, Crowden finishes in a steep rock tumble, yet although there are a couple of pitches which are not really walkers' territory retracing a step or two will always reveal a safe onwards passage. Then again, before reaching the rocky stretch, it is quite feasible to break off from the path and tackle the grassy slope to the plateau where the crash occurred.

Once at the plateau site, and after a period for reflection, the scramble up to the rim path is easily accomplished.

7

Avro Anson Mk.1 N9853

Edale Moor, Kinder Scout

SK 10123 87866	618 m
Unit and Station	No. 16 (Polish) Service Flying Training School, RAF Newton, No. 21 Group, Flying Training Command
Date	11 December 1944
Crew	Two crew, three passengers, Polish Air Force under British Command, all injured Flight Lieutenant Aleksander Chełstowski, pilot Flight Sergeant Stefan Pasiński, wireless operator Flight Lieutenant Melcinski, flying instructor, passenger Flight Lieutenant Witold Siuda, flying instructor, passenger Flying Officer Jan Klimczak, electrical engineering officer, passenger

Although designed as a maritime reconnaissance aircraft, the Anson was to serve as a crew trainer for much of its long and successful life. While generally well liked there is no doubt that it had its drawbacks, even as a crew trainer. It had virtually no performance on one engine, it was noisy, draughty and cold, and if the rubber 'in-flight relief' tube had been kinked by a previous user it could be rather 'orrid to boot. Nor was the task of manually operating the undercarriage on the earlier versions anything to write home about, for it had to be cranked some 160 times! In its favour, however, the Anson was sturdy and dependable and with its 158 mph (138 knots) cruising speed it proved equally valuable as a communications aircraft.

On 11 December 1944, when Flight Lieutenant Aleksander Chełstowski and wireless operator Flight Sergeant Stefan Pasiński got airborne from RAF Newton (east of Nottingham) in Anson N9853, they had been detailed to take two Newton-based Polish flying instructors to RAF Millom, near Broughton-in-Furness, Cumbria. Their third passenger was Newton's electrical engineering officer, Flying Officer Jan Klimczak, who was to carry out an inventory check on arrival. All on board were members of the Polish Air Force under British Command.

Crashed 11 December 1944

***Avro Anson N9853* (MOD Polish archives)**

Flight Lieutenant Chełstowski had been advised of snow-bearing cloud massing along the route above the hills, but only twelve minutes after getting airborne he found himself already in cloud and flying on instruments. With 800 hours' total flying time under his belt, and 137 on Ansons, he deduced that the weather deterioration had occurred more quickly than expected and decided to descend from his cruising height to continue contact flight beneath the cloud. In fact, the deterioration had been so rapid that the cloud was now draped upon the surface. Before he could appreciate this, however, and although his altimeter showed what he considered an adequate clearance, Flight Lieutenant Chełstowski had flown his aircraft into the snowy, deeply-rutted moorland top of the Kinder Scout plateau at an altitude of 2,030 feet above sea level. Moreover, despite the general flatness of the terrain it was not a soft impact, the aircraft rolling onto its back and breaking up.

Gradually collecting themselves the shaken, but only slightly injured, Flight Lieutenants Melcinski and Siuda were able to extricate their bewildered pilot by way of the shattered windscreen. Flying Officer Klimczak, who had been seated beside the pilot, they found to be trapped in the wreckage. Further, having injured both arms and a leg, he was periodically lapsing into unconsciousness. Wireless operator Pasiński, for his part, had suffered some seemingly severe trauma to his body. All they could do, however, was keep their more seriously injured comrades warm by wrapping them in parachutes.

Had they known it they were virtually on track, but as the weather conditions ruled out search aircraft being able to contact them in any reasonable time, Flight Lieutenant Melcinski elected to go for help. It must have been a daunting undertaking, as even a fine-weather survey of the crash-site terrain shows. As it was, still in a state of shock, he blundered in a south-easterly direction, found and followed a watercourse, descended Grindsbrook Clough and duly raised the alarm in Edale. A noble effort! Even so it was early afternoon before rescuers began to arrive at the scene.

It was found that the pilot was more dazed than injured but that the wireless operator had suffered liver damage. It was also established that Flying Officer Klimczak had, indeed, broken both his arms and a leg. His barely conscious state, however, undoubtedly eased his passage down to Edale. In fact, it was to be many days before he fully recovered consciousness and he would only return to duty after nearly two years of surgery and recuperation; notwithstanding which he served on for another two years before finally taking a disability discharge.

Even before the court of inquiry had begun its deliberations Flight Lieutenant Chełstowski's commanding officer had rounded upon him, declaring that, in his opinion, his subordinate should no longer fly as a pilot. And with rather less clouded judgement the members of the court of inquiry were quite unequivocal in their condemnation of Flight Lieutenant Chełstowski's conduct of the flight. They found that in conditions of poor visibility he had descended his aircraft from a safe height and destroyed it, injuring some of his passengers and crew in the process. Further, they discovered such lapses in the pre-planning of the flight that the Air Officer Commanding (AOC) ruled that 'pre-flight preparations were haphazard and inadequate and deserving of severe censure'. The ultimate outcome was that Flight Lieutenant Chełstowski continued to fly, but with the daunting endorsement of 'Gross Negligence' entered into his flying log book.

In passing, it is noted that the AOC held that, being uncertain of his position, the pilot 'should have been flying at least 1,000 feet above the highest ground on route as he

The mid-gully site

could not see the ground'. Which shows that in that era the aircraft would have been considered safe had it maintained 3,088 feet on the altimeter, Kinder's summit being 2,088 feet above sea level. In comparison, the 'safety height' from the early fifties onwards would have demanded a minimum altimeter reading of 3,800 feet, affording a far larger margin to accommodate navigational errors but also to compensate for errors associated with the pressure altimeter: its acceptable limits of accuracy, the mis-setting of its subscale, and even a degree of misreading.

Flight Lieutenant Aleksander Chełstowski, despite the brash condemnation by his commanding officer, flew on and finished the war as the holder of the Polish Cross of Valour, awarded three times over, and the fifth class of the highest of all Polish military decorations, the *Virtuti Militari*: the equivalent of a Victoria Cross.

Of the other occupants of Anson N9853, Flight Lieutenant Witold Siuda also earned the coveted *Virtuti Militari*, Class Five, together with three bars to a Cross of Valour and a Silver Cross of Merit; while Flight Sergeant Stefan Pasiński, promoted to warrant officer, left the Service bearing a bar to his Polish Cross for Valour.

Visiting the Site

From the head of Grindsbrook Clough the route to the Anson crash site follows the rim path northwards for some 400 yards until it rounds the main Grinds Brook tributary at SK 10610 87600. At this point it continues upstream along a broad, shallow water channel which affords easy going, the walker switching from rough path to stream bed as it suits. After some ten minutes (another 400 yards), at SK 10433 87932, a gully opens off to the left. This immediately forks, the right fork (initially bearing 290°M) running westwards for another ten minutes (yet another 400 yard stretch) before encountering the wreckage.

The iron-man might prefer to reach the site by stepping up to the singular double-rock feature where the rim path kinks at Grinds Brook, then striding over heather and drainage channels for 600 yards on a heading of 308°M. The timing, though, will hardly alter.

Back in 1944, the recovery team tumbled the remains into a convenient gully; in part because recovery was so difficult from such a remote site, in part to prevent the wreckage from misleading any future aerial searches. Indeed, the area is so broken with groughs – as the region's minor water gullies are known – that although a moderate amount of wreckage remained in early 2010, it may well be necessary to search around from even the best of locational references in order to discover the correct gully.

The most straightforward return from the Anson site is to follow its debris-strewn gully back to join the Grinds Brook tributary and so the rim path and the head of Grindsbrook, a half-mile, 25 minute, walk.

It would be most unusual, however, to visit the Anson site in isolation, the crash site of Rapide G-ALBC (see below) being just 400 yards away – but some 20 minutes over such rough ground. In fact, from the furthest extent of the Anson debris the Rapide site bears 010°M, a good aiming point being a prominence (623 m on the map) necklaced by boulders, the Rapide having come down at its base. Because the drainage runs at right angles to the route, and with the terrain being badly broken, the initial part of this excursion means scrambling into and out of several groughs. Beyond the wide swathe of the Grinds Brook tributary, however, the going becomes easier.

De Havilland DH89 Rapide G-ALBC

Edale Moor, Kinder

SK 10160 88241	607 m
Operator	Solair Flying Services, Birmingham
Date	30 December 1963
Crew	Two, both injured Captain Dennis Holmes Co-pilot John McWhirter

Although only 700 de Havilland DH89 Dragon Rapides were to be built, and although production ceased in 1946, the 1934 transport was to become one of the United Kingdom's most cherished designs. Many were commandeered by the RAF on the outbreak of the Second World War, when, renamed the Dominie, they were used for training and communications purposes. However, for years after that the Rapide's tapered wings, streamlined but fixed undercarriage, and its by-then singular biplane configuration, made it a familiar novelty in the sky.

With its light construction and easy adaptability, whether for freighting, passenger-flying or parachute dropping, the Rapide was understandably popular with emergent air operators, among them the Birmingham-based Solair Flying Services, which used Rapide G-ALBC for photographic surveys.

On 30 December 1963, having completed their assigned task, a Solair crew, Captain Dennis Holmes and his co-pilot, Mr John McWhirter, were returning to Birmingham from Middleton St George, a wartime RAF airfield and for many years since Teesside Airport. For some reason, as he approached halfway along his 140-mile course, Captain Holmes decided that it would be politic to divert to Manchester and refuel. However, while his direct routing to Birmingham would have taken him well to the east of Ladybower and over relatively low-lying terrain, his change of course to Manchester took him skimming over the crag-rimmed plateau of Kinder Scout, an expanse of moorland raised to an average height of some 2,000 feet above sea level.

From a weather pattern giving high winds and billowing cloud, strong vertical gusts obtained over the plateau. When a downdraught embraced the presently ground-hugging machine, it effectively sucked it from the sky at such a rate that its twin 200-horsepower Gipsy Six engines were quite unable to counteract the descent.

The wreckage as the salvage team found it

The machine struck heavily and disintegrated, but providentially there was no fire, for both crew members were incapacitated. Co-pilot McWhirter suffered head injuries which, while painful, eventually turned out to be slight. Captain Holmes, on the other hand, sustained a broken leg which immobilised him amidst the wreckage.

The downdraught had struck too swiftly to permit a distress call, so both knew from the outset that they were in for a cold and uncertain wait. In fact, it was to be colder and more uncertain than they might have anticipated, for although they were spotted by a helicopter before darkness fell, conditions had so worsened that the helicopter crew were unable to even contemplate landing. It was to be the fortunate chance of a glint of moonlight on a metal panel that eventually led a blizzard-bucking mountain rescue team to the site. Even then it was a matter of leaden-footed hours before Mr McWhirter was helped, and Captain Holmes carried, off Edale Moor and down to where proper care could be given.

Crashed 30 December 1963

One of the first on the scene had been Mr Gordon Miller, by title 'Area Ranger for Kinder – including Edale, Castleton, and Hayfield', who recalled that the first thing both pilots asked about, and continued to be concerned over above all else, was the safety of the camera and the films; that only when assured that these had been secured would they turn to their own requirements.

The crash site, looking towards the prominence

What Solair thought of its captain being caught out by a downdraught is not recorded, but there was nothing to be done with the remains of the stately old Rapide beyond incinerating what would burn in order to prevent its distracting any future airborne searches. Indeed, Area Ranger Miller spoke graphically of the bonfire, photographed by his associate, Peak Park Ranger Jim Buie, and particularly of 'the writhing iridescence' as petrol-soaked, painted wood and doped fabric flared up in the breeze.

Because of the relative remoteness of the crash site there was still a substantial amount of wreckage at the scene in 2010, notably the cylinder block of one of the engines, but little else identifiable; just a general scattering of intermingled wood and metal fragments. And little enough beyond the site. Nothing but the nearby lift of boulders above a moorland otherwise largely featureless and deeply rutted to any horizon the uncertain Kinder weather may allow by a myriad twisting groughs and hags.

Visiting the Site

From the head of Grindsbrook Clough the route follows the rim path northwards for 400 yards until it kinks at the Grinds Brook tributary at SK 10610 87600. From here it continues upstream, leaving the walker to switch from rough path to stream bed as it suits. The going is pleasant, along a wide, shallow grough, and passes a drystone shelter at SK 10362 87998. This has come into being on the site of the former Four Jacks Shooting Cabin – built by four chappies, each named Jack. After half a mile (some 20 minutes over such easy ground), the route angles off the tributary (at somewhere like SK 10080 88150) to head through the hummocks for five minutes towards a hillock necklaced with white boulders – or slate grey, in dull light – marked 623 m on the map. The crash site is in the broken ground at the foot of this prominence.

The most straightforward return route from the site is accomplished by heading for and then following the Grinds Brook tributary back to the head of Grindsbrook Clough, a half-mile, 25 minute, walk.

Having come this far off the beaten track, however, it would be unusual to return without having visited the crash site of Anson 9853 (see above), just 400 yards – but 20 minutes – distant. It lies on a heading of 195°M, beyond the flat expanse of the Grinds Brook tributary and thence over rising and very broken ground. Because the direction of travel runs at right angles to the drainage, several substantial peat groughs have to be negotiated. Even then it may be necessary to cast around to find the actual gully, although the Anson's wreckage extends for some 300 yards.

Despite being just 400 yards from the Grinds Brook tributary, the terrain imparts a sense of just how disorienting these upper moorlands can be in poor visibility, fretted as they are by hags and groughs – heathery tops and water channels – and offering relatively few prominent features.

9

Handley Page Heyford K6875

Broadlee-Bank Tor, Edale

SK 11092 86013	530 m
Unit and Station	No. 166 Squadron, RAF Leconfield, No. 4 Group, Bomber Command
Date	22 July 1937
Crew	Six, all killed Sergeant Newton W. Baker, pilot Sergeant C.P.D. McMillan, second-pilot Sergeant Jim W. Barker, pilot (navigating) Aircraftman First Class Harold Gray, wireless operator Aircraftman First Class Eric McDonald Aircraftman First Class E.J. Musker

The 1933 Handley Page Heyford was an all-metal biplane-bomber whose speedy 143 mph (124 knots) earned it the appellation 'Express'. Although withdrawn from first-line service in 1939 the type still gave good value as a crew trainer until 1941, being stable and pleasant to fly. But like all aircraft it needed airspace, and when this was denied it the results could be catastrophic; as they were for the occupants of No. 166 Squadron's K6875 on 22 July 1937.

The aircraft was being operated by Sergeant Pilot Newton Baker and his crew, the six airmen having been dispatched from RAF Boscombe Down to carry out a night navigational exercise while returning to RAF Leconfield, their home station, near Hull. In the course of the flight, however, they became lost, strayed 13 miles off track, and crashed into the high ground of Broadlee-Bank Tor, Edale, dying to a man.

The aircraft had been seen to cross Rushup Edge and fly on up the heavily clouded Edale Valley towards Kinder at not much more than a thousand feet above the valley bottom, a perilously low altitude for that region; further, it was a cloud-darkened night with the summits in mist and with frequent rain showers cutting down visibility still more. Providentially, it was clear that the crew had no idea that they were in any danger, for as investigator Squadron Leader Hugh Wake found, 'having interviewed the most reliable witnesses…the engines were running normally at the time of the accident. It [the aeroplane] did not circle round or fire any lights, and was not in trouble.'

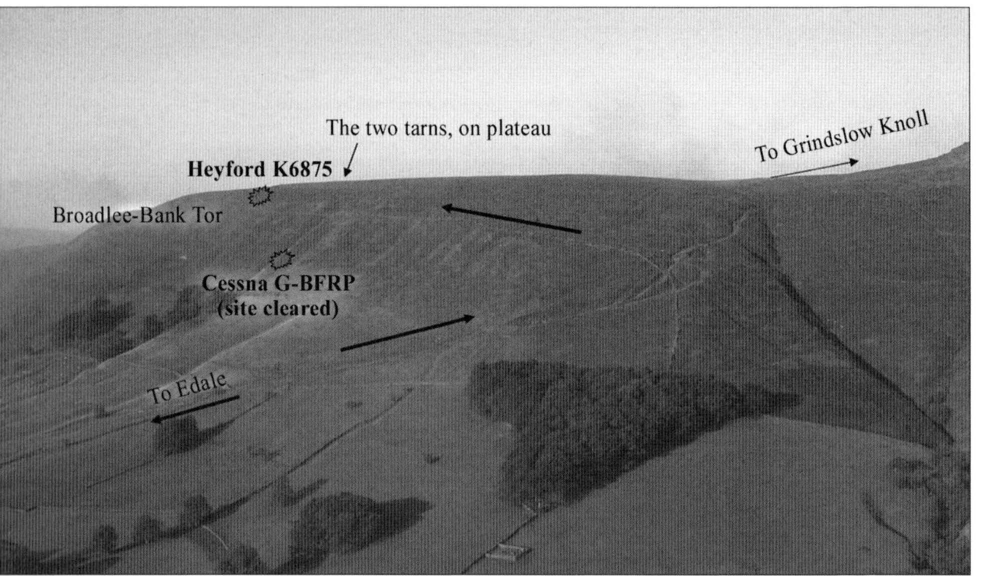

The route followed by the would-be rescuers; also showing the upland route to the west of Grindsbrook Clough

On the fateful night itself, the outcome had not been long delayed for, as if inevitably, the moment came when the starboard wing clipped the shoulder of the 1,700 feet above sea level Broadlee-Bank Tor, yawing the Heyford fierily into the ground. Those who had watched in trepidation swiftly laboured up the zigzag shooters' path to the site, only to discover that the aircraft had smashed through and levelled 50 feet of a drystone wall, and that the crew were beyond all aid. 'The summit is in thick mist,' Mr H.W. Porter of the Nag's Head Inn told the authorities on the phone, 'and the wreckage is still burning. We found two bodies near the aircraft, and four inside it, but nobody could have lived in that blaze.'

The court of inquiry had little to do. As Squadron Leader Wake had ascertained, the aircraft had shown no sign of being lost, and imaginative early newspaper reports – and accounts derived from them – notwithstanding, had neither circled nor sought to determine its position with flares. It was clear, then, that the crew had not been in the least perturbed; that quite unaware of the fact that they were 13 miles off track and therefore flying below the level of high ground, they had simply been going about their in-flight duties, chatting idly no doubt, while waiting out the lapse time to the next turning point, the unwonted, catastrophic impact taking them totally – and blessedly – unawares.

Crashed 22 July 1937

The accident had clearly been avoidable and the finding of the court of inquiry was correspondingly brief: navigational error. The court further observed that the aircraft had been flying at an altitude which made no allowance for the sort of navigational error that had, in fact, occurred. Both cause and effect being clear cut, nothing else needed saying.

Nevertheless, shortly after the court had delivered its findings, Squadron Leader Wake wrote a personal letter to Sergeant Jim Barker's widow, Mrs Muriel Barker, a letter relied upon throughout this narrative by the courtesy of Mr James Watson, of Tunbridge Wells, a relative of Sergeant Barker. In his letter Squadron Leader Wake expressed the fellow feeling that all members of the inquiry would have experienced: 'I blame no one for the accident which was due solely to the aircraft being slightly off its course and over high ground. Had it been on its course it would have been clear of the hills. This slight error could easily occur in conditions of low cloud, and, as we know well, happens frequently to all of us.' How true!

It is not known if Squadron Leader Wake wrote a similar letter to all the bereaved families, but one has the feeling that he singled out Mrs Barker with a special consolatory purpose. For while it has to be admitted that the disposition of crew tasks on this particular flight has not been positively established, the function of two of the

The debris pool in early 2010, looking back along the line of impact, showing the drystone wall the Heyford crashed through

three pilots on board was clear: Sergeants Baker and McMillan were operating as first and second pilot respectively. Aircraftman First Class Harold Gray was the wireless operator, and it is a fair assumption that Aircraftmen First Class Eric McDonald and Musker were either gunners or mechanics. Which leaves qualified pilot Sergeant Jim Barker to be acting as navigator. Certainly, the crew disposition would have been common knowledge on the squadron – and most certainly in married quarters – at the time. Therefore, one notes the gentleness with which Squadron Leader Wake reiterates that the error was slight, and one which not only could happen to any of them, but one which 'happens frequently to all of us'.

Setting aside both responsibility and tragedy, even the composition of this early crew is of interest, for at that time aircrew categories were not as well established as they would later become, and airmen fliers were essentially ground tradesmen and only part-time fliers. Even the captain, Sergeant Baker, a pilot of some eleven years' standing, had only recently returned to flying duties after five years back in his ground trade.

Regarding the crash site of the Heyford, Mr Robert Allen Atkin, of Lady Booth Farm, Edale, described how he helped bring down the wreckage from the scene. 'At the time,' he explained, 'Maurice Oaks, whose land it was, had a Fordson Standard tractor, but he couldn't get all the way up with it, so we took a pair of horses and dragged the wreckage down to him.'

Visiting the Site

Visiting the Heyford site requires a one hour twenty minute (one and a half mile) walk, embracing 1,000 feet of ascent, from Edale public car park. The least strenuous route initially takes the Pennine Way footpath leading westwards from Edale. After 400 yards it leaves the Pennine Way and branches right as an ascending grassland track which soon develops into a well defined, deeply grooved shooters' trail. The trail itself, climbing towards Grindslow Knoll, keeps high on the south-western shoulder of Grindsbrook Clough, so offering splendid overviews of that justifiably popular pathway. However, at SK 11580 86445, a hard left turn is made to follow the drystone wall climbing south-westwards to reach the Heyford crash site.

As it happens, in following the wall the way passes above the bilberry-covered slope from which every trace of the wreckage of Cessna G-BFRP (see Non-Debris section– page 143) was cleared.

At the Heyford crash site what little wreckage remained in early 2010 had years since been gathered into a rock circle just north of the wall the aircraft had crashed through. The sparsity of debris aside however, what the site certainly offers – on any good day! – is a stunning panorama of both the Mam Tor range and the Kinder massif.

From the Heyford crash site a reasonable track (ascending, in fact, from Upper Booth) leads northwards, leaving two tarns to the right and heading towards Grindslow Knoll and so to the rim path running around Kinder.

10

Vickers Armstrong Wellington Mk.1C W5719

West Upper Tor, Kinder Scout

SK 11061 87550	567 m
Unit and Station	No. 150 Squadron, RAF Snaith, No.1 Group, Bomber Command
Date	31 July 1941
Crew	Five killed, one injured Sergeant Percival Harold Charles Parrott, pilot Sergeant Joseph Arthur Haswell, supernumerary (second) pilot, on operational experience Sergeant Jack Douglas Evelle, observer (navigator), Royal Canadian Air Force Sergeant Frederick Kenneth Webber, wireless operator/air gunner Sergeant Dennis Aloysius Monk, air gunner Sergeant Earl Tilley, air gunner, tail turret (injured)

In the early days of the Second World War the twin-engined, medium-range bombers carried on what would later be seen as a merely embryo offensive against Germany; the Whitleys by night, and the Hampdens, Blenheims and Wellingtons by day. But in mid-July 1941 they were the only force capable of carrying the fight to the enemy and accordingly, despite all the shortcomings the outbreak of 'the shooting war' had revealed, it was left to them to persevere in the task. But it was not a wasted effort, for their crews were learning the lessons, gaining the experience, and in truth sowing the seeds which would spring up, mature, and ultimately be reaped as a vastly more destructive harvest in the campaigns yet to come.

Among the shortcomings revealed was the near impossibility of accurately navigating at night over a hostile, blacked-out continent, with no effective radio aids to assist, and no bomb-aiming gear effective enough to make the raid count in the event of actually reaching the target. There was also the ever-increasing effectiveness of the enemy's air defences. Not forgetting a weather pattern that would as often as not baulk the raiders

The terrain the crew descended into, the sunlit impact point of Wellington W5719, and the flat moor above

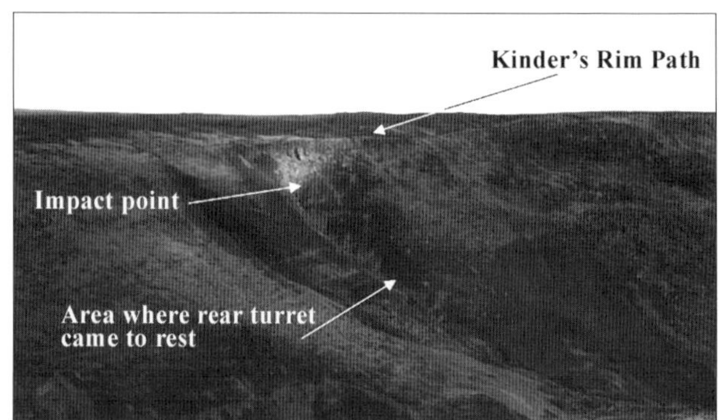

in locating the target, and then treat them even more harshly when they sought to return to their airfields in the low-lying, mist-prone eastern areas of the United Kingdom.

Again, in the early days in particular, it was held to be of paramount importance to bring the bomb load back if there was no reasonable certainty of hitting the designated target; anything to avoid the charge that the RAF had bombed a non-military locale. But this humanitarian precept had long gone by the board when, on the night of 31 July 1941, Sergeant Pilot Percival Parrot and his crew, together with seven other Wellington crews from RAF Snaith, were briefed to raid Cologne.

The debris pool in 2010, looking towards Edale, with the path along Grindsbrook Clough below

The weather on the outbound leg over the North Sea was cloudy, but beyond the enemy coast thunderstorms developed which eventually proved so difficult to circumvent that, in the end, Sergeant Parrott made the decision to abort the operation and turn for home, opting at the same time against jettisoning his bomb load into the sea. Humanitarian precepts aside, bombs were expensive!

Having made the turnabout from an arbitrary position, and with thick cloud blotting out both ground features and the stars, Sergeant Evelle, the Canadian navigator, returned to his air plot (see Glossary), but with little to go on beyond dead (deduced) reckoning. Later, however, he felt confident enough to advise the crew that they had re-crossed the coast in the vicinity of the Wash, and had just 75 miles to go.

Despite static interference on the radio, Sergeant Webber, the wireless operator, managed to raise base and obtain a magnetic course to steer for Snaith. At which stage it must have seemed as if Sergeant Parrott could relax a little, for now it became a case of simply holding the given course until the overhead-time calculated by the navigator. Not that Sergeant Parrott could relinquish the controls, for the supernumerary pilot was only aboard to gain experience prior to being tasked for operations with his own crew. Moreover, any relaxation was to be short-lived, for as they neared the estimated arrival time for Snaith, they still had total cloud cover below them.

So it was that, anticipating that they were at least in the vicinity of low-lying Snaith, a descent through cloud was begun. Only, with their altimeter still reading 2,000 feet – so persuading them that they still had some 1,700 feet of clear air below them – the aircraft impacted into the 2,000 feet above sea level crags of Upper Tor, on the southern rim of the Kinder massif, and some 37 miles south-west of Snaith.

The bomb load exploded on impact, killing everyone in the forward part of the aircraft. But the impact had also jarred loose the rear turret, which then broke off and bounced down the precipitous slope below the crags, carrying air gunner Sergeant Earl Tilley clear of the explosion. Hard-proven statistics had it that the rear-gunner's station was the most vulnerable in the aircraft on operations and yet its very vulnerability paid off here.

Having collected himself, Sergeant Tilley, realising that he had suffered only minor injuries, was able to make his way down the two-mile length of Grindsbrook Clough to Grindsbrook Booth (effectively, Edale), where the alarm was raised.

The subsequent investigation would find that, although the pilot had accurately flown the heading passed over the radio, he had made no attempt to re-check the validity of the navigator's estimate for the Snaith overhead before descending blindly through cloud: the corollary being that neither the pilot nor the navigator was ever to realise that the aircraft had actually overflown Snaith and then carried on towards the high ground. Continuing the accusative theme, higher authority caustically observed that none of several laid down lost-procedures had been employed, and that no emergency get-'em-home service had been called upon for help. Accordingly the accident was attributed to an error of judgement.

Regarding the rest of the force sent to raid Cologne that night, five aircraft found the city and bombed through cloud, while two, totally thwarted by the weather, found and bombed targets in Belgium. Of these seven, six returned safely to Snaith, with the seventh crashing having crossed the Channel.

As for fortune-favoured Sergeant Tilley, he eventually returned to operational flying, and despite being subsequently shot down and held as a prisoner of war for three years, survived the conflict.

Standing across the valley from the crash site and surveying the horrid aspect which that cloudy July night so considerately withheld from the hapless crew, the imagination reels. Above, by just a matter of 50 feet, a virtually level plain; below, steep slopes falling dramatically to silvery Grinds Brook far beneath. But at the point of impact itself, sheer unyielding gritstone crags.

Visiting the Site

Probably the most interesting route to follow in visiting the crash site of Wellington W5719 begins with the popular 1,164 foot, 2 mile ascent up Grindsbrook Clough. The path is well established but, as with most cloughs, the final ascent is through a steeply-rising rock field. This is less daunting than it appears, however, offering a myriad ways up and taking just ten minutes to scale. At least, that was the time assorted walkers took while being observed during a break for salted peanut butter sandwiches and hot blackcurrant juice.

From the head of Grindsbrook Clough the route follows the rim path for about 15 minutes, first northwards around Grinds Brook as it feeds into Grindsbrook Clough, then eastwards to the crag at the western end of Upper Tor into which the aircraft crashed.

The debris pool can be seen without leaving the rim path by stepping aside and peering down from the flat top of the crag. Actual access, however, calls for a very careful, scrambling descent to the east of the crag, having accomplished which a tiny memorial tablet will be found on a sheltering rock above the debris pool.

The crash sites of Anson N9853 and Rapide G-ALBC make an interesting addition to this foray (see above). To do this it is necessary to return to the Grinds Brook tributary (at SK 10610 87600) then follow it upstream for approximately 30 minutes.

A fine return route to Edale is initiated by backtracking to the head of Grindsbrook Clough, holding the height, and walking south-eastwards towards Grindslow Knoll. Indeed, it is on the approach to the Knoll that the prospect across the valley towards Upper Tor shows just what the hapless Wellington crew were descending into! And how just a few feet higher might have meant impacting into flatter, and conceivably kindlier, moorland. But then, as in every such case, just for a matter of ...

Having either scaled or skirted the Knoll, a direct descent can by made by a shooters' path. Alternatively, a slight extension can be made to take in the crash site of Heyford K6875 on Broadlee-Bank Tor (see above). To visit the Heyford site it is necessary to branch right from the shooters' path at the foot of the Knoll and walk a few hundred yards off-track aiming to the right of a pair of tarns. This will connect with a path (eventually bound for Upper Booth). The drystone wall seen on the skyline lies just beyond the Heyford crash site.

From the Heyford site the least demanding descent to Edale follows the rough track alongside the drystone wall. This runs north-eastwards until it intersects with the broad path leading from Grindslow Knoll. In its passage the wallside route leads above the steep, bilberry-covered slope from which the wreckage of Cessna G-BFRP was cleared (see the Non-Debris section– page 143).

11

Vickers Armstrong Wellington Mk.3 X3348

South-west of the 590 m Kinder-East trig column

SK 12803 87554	590 m
Unit and Station	No. 427 Squadron, Royal Canadian Air Force (RCAF), RAF Croft, No. 6 Group, Bomber Command
Date	26 January 1943
Crew	Six, all injured Pilot Officer Carl A. Taylor, RCAF, pilot Sergeant G.T. Southwood, RCAF, supernumerary (second) pilot, on operational experience Pilot Officer D. Martin, RCAF, navigator Pilot Officer Donald. R. Mortimer, RCAF, bomb aimer Sergeant A.P. Deane, RCAF, wireless operator Sergeant W. Lumsden, air gunner

By November 1942, when No. 427 Squadron of the Royal Canadian Air Force (RCAF) formed at RAF Croft, just south of Darlington, and received its Wellingtons, the type had already been diverted to night, rather than day, bombing. Further, its performance and bomb load were beginning to show palely in comparison to those of the four-engined Halifaxes and Lancasters being introduced into service; in fact, No. 427 itself would re-equip with Halifaxes in May 1943. In the six months before that, however, the squadron used its Bristol Hercules-engined Wellington Mk.3s to great effect, not least on 26 January 1943, when nine of its aircraft, Wellington X3348 among them, attacked the U-boat pens in German-occupied Lorient, on the west coast of France.

It was a marathon trip to Lorient, on the far side of the Brittany peninsula, and even on the return, coasting-in over Dorset's Lyme Bay, the raiders were still faced with another 300-mile flight back to their County Durham base; a flight affording time and distance to spare in which to wander off course and either fall behind or race ahead of estimates, with only a forecast wind many hours old to help in calculating their progress. But the basis of crew co-operation is to have faith in one another's expertise, and the navigator's station, where Pilot Officer Martin hunched over his lamplit plotting chart, became the natural depository for that faith.

In this case the navigator was assisted by the bomb-aimer reporting any airfield pundit-beacons opportunely spotted from the nose; pundits being lights flashing a two-letter airfield identifier in morse. Except that one of these, glimpsed through cloud, proved contentious; possibly because of cloud intervening but equally likely because a pundit's painfully slow coding invariably has to be watched through at least one full cycle before it is possible to determine which are dots and which dashes.

In this case, when the perverse identifier was finally agreed upon, it showed them to be considerably further along track than Pilot Officer Martin had previously reckoned on. Nevertheless, accepting the doubtful offering of this so-critical visual fix – and doing so a shade too casually, it is hazarded – he calculated a new heading for RAF Croft. Just a little later he assured his pilot that it was now safe to descend through cloud on that heading.

But pilots (being sensitive souls) get 'feelings' sometimes, and when Pilot Officer Carl Taylor, an American serving with the RCAF, was well into the descent he suddenly decided to put on power and regain some height. The same discomfort, it seems, had also made him edge slightly eastwards from the new heading; although it is possible that when the navigator passed the adjusted heading, the supernumerary pilot, standing between him and the captain, and on board to gain operational experience, inadvertently set the wrong figure against the compass lubber line. Not an unknown mistake to make, particularly when tired, stressed, and working in the poor cockpit-lighting of a night-wrapped bomber. Indeed the passage from the contemporary *Air Navigation* manual for the RAF, quoted elsewhere, is relevant once again: 'The Navigator must keep a wary eye on the...Pilot, who, for various reasons, may not be steering the requisite Course. It is advisable to look repeatedly over the Pilot's shoulder at the Course being steered.' Although this sort of overseeing might well have risked denting the aforementioned perfect faith.

Whatever the truth about the heading passed and steered, it is a matter of record that after six hours and 40 minutes of flight, and by the grace of God with some height regained and therefore with some 2,000 feet now showing on the altimeter, the aircraft abruptly lost flying speed as it unaccountably made contact, then bellied its way across the ground, eventually coming to a halt, battered but on an even keel and without catching fire.

In October 2009, Mrs Milly Heardman, of Edale, then 'ninety-five and a half, not ninety-six!', was able to describe the way the aircraft had appeared. 'As you approached it looked quite whole, as if it could fly away at any moment. The wings were spread out, and there didn't seem anything wrong with it; but when you got close, you realised that the whole of the bottom was smashed.' Mrs Heardman's late husband, a Moorland man of large repute – Heardman's Plantation, passed on the way to the site, is named for him! – attended this crash, as he did so many. As Milly

Crashed 26 January 1943

The debris pool of Wellington X3348 in early 2010
Inset: Mrs Milly Heardman, of Edale, October 2009

said fondly, 'Fred, that was the name over the doors of both the Edale pubs he came to own, but he was Bill, or "Bloody Bill, the bog trotter", otherwise.'

In the aftermath of the touchdown, the shocked crew had scrambled free, found that none of them had any serious injuries, and discovered that they were on high moorland rather than on the low-lying pastures they might have expected in the vicinity of Croft. They also realised that the moor was exposed, that rain was falling steadily from the overcast, and that the night was chill. At which they re-entered the largely intact, if somewhat tattered, fuselage and settled down to wait out the remaining hours of darkness.

But as time dragged on the desire to inform the squadron that they were not, in fact, 'missing', proved more powerful than their patience, and three of them set out to seek help, carrying with them a torch and the aircraft's Very signal pistol (see Glossary) with a supply of cartridge flares.

The going was rough, over trackless, gully-rifted heather bog, but at a critical moment they had a remarkable stroke of luck. For initially attracted by a sound from far below, they saw, piercing the pitch-dark opaqueness of a blacked-out land, a red glare, moving horizontally and snuffed out even as they watched. Afterwards they would reason that it was the briefly opened firebox of a train passing along the Vale of Edale on the Sheffield to Manchester line. But for the time being it was enough to know that there really was life of some sort down there in the gloom. Reinvigorated, they stumbled onward, resorting to firing off flares when uncertain of the footing, to eventually find succour on the outskirts of Edale; at which juncture they would discover that they had been just 12 miles east of track, but a full 65 miles short of Croft.

The subsequent court of inquiry worried away at both errors earlier suggested: that the pilot may have ignored the navigator's heading, and that the supernumerary pilot may have mis-set it. In the event, however, the final submission was that 'bad pilotage' (see Glossary) had caused the loss of the aircraft; a submission with which higher authority simply concurred. After all, pragmatically viewed, the aim of the operation had been fulfilled, Lorient had been bombed, and all the other aircraft had returned safely, albeit diverting to airfields other than their base with fuel states running low.

And nothing was said about the causative misidentification of the pundit beacon: crew loyalty triumphing, one suspects!

Visiting the Site

The route to the Wellington site from Edale public car park requires an 80 minute, 2 mile climb, making the height of 360 metres (1,200 feet) on well-defined upland paths. Initially the route leads north-eastwards, away from the paved Grindsbrook Clough path and up The Nab, this zig-zag section being the steepest of the trek!

Beyond The Nab, three paths offer, but although seemingly steep, the central one – not shown on one-inch maps, but obvious on the ground – offers a fine and unchallenging rocky route up the very spine of the Ringing Roger crag. This leads in a direct line to the Wellington site with the ascent initially flattening onto more open ground but changing swiftly to become very broken with spongy hags and soggy groughs. Because of this, with the debris pool being minimal, searching around will probably be required no matter that the GPS reading given here is so well proven.

This sortie will almost undoubtedly take in the nearby Halifax HR727 site (see below), and the Kinder-East column. The Halifax site is spaced across the line of travel from the Wellington site, so to reach its impact point head about 076°M and about 055°M (see Glossary) for its terminal point. And for either, anticipate six or seven minutes of very rough and boggy going. Reflecting, also, that justifiably popular as walking off-path over the Kinder plateau has become, when the mist descends it can impart an awesome sense of isolation.

Having returned to the rim path, however, very scenic descents are offered by both The Nab and Golden Clough. For those extending the walk, 30 minutes (a mile) will bring them past the crash site of Wellington W5719 (see above) and a further 15 minutes to the head of Grindsbrook Clough. An alternative descent then, is via Grindslow Knoll, which can be adapted to take in the crash sites of Heyford K6875 (above) and Cessna G-BFRP (see page 143). Of these two descents, the Grindsbrook route leads down a splendid valley while the Grindslow Knoll route affords spectacular panoramic views.

12

Handley Page Halifax Mk.2 HR727

Near the 590 m Kinder-East trig column

SK 13082 87632	585 m, impact point
SK 13039 87745	586 m, terminal point
Unit and Station	No. 51 Squadron, RAF Snaith, No. 4 Group, Bomber Command
Date	5 October 1943
Crew	Seven; five killed, two injured Sergeant Ernest Hatfield Fenning, RAF, pilot Warrant Officer Class Two, Jean Gilbert Felix Fortin, RCAF, navigator Sergeant Eric George Lane, RAF Volunteer Reserve, flight engineer Sergeant Frank Squibbs, RAFVR, wireless operator Sergeant Boris Carl Short, RAFVR, mid-upper gunner Sergeant Victor Garland, RAF, bomb aimer (injured) Sergeant Jimmy Mack, RAF, rear gunner (injured)

The Halifax was the second of Britain's four-engined heavy bombers, entering service in November 1940, after the Stirling but a year before the Lancaster. Although it has always lived in the shadow of the Lancaster, the Halifax showed itself to be an extremely versatile aircraft, variously serving as transport, glider-tug, maritime reconnaissance, and clandestine delivery vehicle for both the dropping and the landing of agents and supplies. However, when Frankfurt-bound Halifax HR727 took off from its base at RAF Snaith, near Selby, on the night of 4 October 1943 it was operating in its design role of heavy bomber.

The effectiveness of the crew's bombing is unknown, but navigator Warrant Officer Fortin would have been aided in guiding bomb-aimer Sergeant Victor Garland to the target either by his own radar display, or by target-marking flares dropped from a pathfinding bomber using such a radar.

Sergeant Pilot Ernest Fenning and his crew had just bombed their target when they were coned – hedged about – by searchlights and then attacked by a night-fighter, their left-inner engine sustaining damage and catching fire. Just the same, after successfully evading their assailant they were able to stop the engine, extinguish the fire and, having feathered the windmilling propeller (that is, turned its blades edge-on into the airstream to minimise the drag), continue flight on three engines.

It was soon discovered, however, that one of their fuel tanks had been holed, after which flight engineer Sergeant George Lane began minimising the wastage by making the damaged tank the priority fuel source for the three remaining engines. Meanwhile the aircraft, now asymmetrically powered, was settled onto the pre-calculated and tactically-designed circuitous return course that would first bring the crew safely over the pre-warned South of England defences, then take them north to their Yorkshire base.

The rough course having been set, the crew were faced with a nice problem of navigation and fuel endurance. Having done his calculations, though, and having liaised with Warrant Officer Jean Fortin, the Canadian navigator, Sergeant Lane was soon able to assure his captain that it would be four hours before they ran out of fuel. It must have been a relief to Sergeant Fenning as he re-trimmed the rudder yet again to ease the strain on his right foot, for by the navigator's estimate they would have landed at Snaith long before that.

The besetting problem thereafter was to update the weather forecast last obtained many hours before, and in particular the winds affecting them. Unhappily, Sergeant Frank Squibbs, the wireless operator, was unable to supply this information, a failure which forced the navigator back upon his air plot, a draughtsmanlike form of navigation – some would say, esoteric – which, by its nature, grows in error the longer it runs.

Halifax HR727 impact point, photographed in 2010

Fortunately, visual fixes were obtained over Beachy Head, and again over Reading, each allowing the air plot to be restarted from a known point and permitting the updating of both headings and estimates for RAF Snaith. Eventually too, having utilised all but 20 gallons in the damaged tank, the flight engineer began restoring the normal fuel feed to the three remaining engines.

Except that at this juncture the crew's problems took a turn for the worse. For with thick cloud now widespread ahead of them, the aircraft's long-range communications equipment failed altogether, the same power loss cutting them off even from the short-range emergency aid available through the listening stations of the 'Darky' homing organisation. Additionally, and unknown to any of them, despite the recent pinpoints furnished to the navigator, they were drifting inexorably to the left of track. What became only too apparent, however, was that the fuel gauges were suddenly seen to be reading unaccountably, and very alarmingly, low.

Veteran Ranger John Campion 'Campy' Barrows.* Inset *July 2006, long retired, but ready to hold weekend court at Fairholmes Ranger Station. Campy died in March 2009.

Now faced with the imminent, and hitherto totally unexpected, threat of running out of fuel, Sergeant Fenning decided upon letting down in a gradual descent, hoping that on breaking cloud they could pinpoint themselves once more. After all, he would have reasoned, the navigator's revised heading from Reading took them well to the east of any high ground, while Snaith itself lay at very nearly sea level. Only abruptly, with some 2,000 feet still reading on its altimeter, and with an indicated cruising speed, enhanced by the descent, of some 250 mph (217 knots), the Halifax impacted heavily into solid ground, breaking apart and instantly killing four of the crew, Sergeant Pilot Fenning among them.

The survivors, Sergeant Jimmy Mack, the rear gunner, and Sergeant Victor Garland, the bomb aimer, gathered concernedly about flight engineer Sergeant Eric Lane, finding him in a bad way. In fact, within a matter of hours Sergeant Lane was to die of internal injuries. But having tried to make him comfortable, and leaving him in the care of Sergeant Garland, whose injured ankles prevented him from walking, Sergeant Mack set out to seek aid.

Blundering his way northwards over waterlogged and hummock-hagged grasses, then down near-precipitous slopes through stygian darkness, it was only when he eventually reached a cottage in the Woodlands Valley and raised the alarm that Sergeant Mack discovered that their Halifax had crashed above Blackden Edge, 1,600 feet up on high Derbyshire moorland, over 1,500 feet above, and a full 35 miles from, their low-lying aerodrome at Snaith.

The joint causes of the accident settled upon by the court of inquiry were lack of fuel and faulty navigation. The flight engineer was singled out for particular blame, both for miscalculating the fuel endurance and for failing to keep a check of the fuel actually being used until the state had become critical; blame upheld by the various senior commanders who declared him primarily responsible. Sergeant Pilot Fenning, for his part, was found to have made an error in captaincy in letting down through cloud when uncertain of his position, and also in guilelessly taking the flight engineer's fuel-endurance estimate on trust. The court had before them, of course, Sergeant Fenning's record of just under 500 hours' total flying experience, woefully low in truth, but normal enough for those days.

Such reproofs may appear harsh, yet the reality that would have been recognised by all aircrew, including the reporting officers (many experienced in, or resting from, operational tours themselves) was that if, like most returnees of the bomber force engaged that night, the crew had brought their damaged aircraft safely back to a base – any base – there would have been acclamations all round for a job well done. As it was, avoidable errors had prevented them from doing so and in consequence Halifax HR727, both valuable and costly, lay a worthless heap upon a Derbyshire peat bog; while dead amidst its wreckage lay five of its crew.

As a footnote to the cost of such bomber accidents, the Air Officer Commanding-in-Chief of Bomber Command, 1942-1945, later to become Marshal of the Royal Air Force, Sir Arthur Harris, is quoted as saying that an aircrew paid for its training and its aircraft if lost after two successful operations, having bombed its target on the third. As an aircrew member observed drily, 'A credit tick, and all for the price of seven telegrams.' For Sergeant Fenning's crew this had been their eleventh operation. The only survivor fit to return to flying, bomb-aimer Sergeant Victor Garland, eventually completed a first tour of 30 operations and then embarked upon a second tour, flying a further seven before the end of hostilities.

Mr Maurice Cotterell, of Hayridge Farm, in the Woodlands Valley, remembered the grim aftermath of the 1943 crash. 'I was at Gillot Hey Farm at the time,' he recalled, 'and the plane hit the top between this side of the hills and Edale. The bodies were brought down to our little chapel.'

Mrs Milly Heardman, of Edale, on the far side of Kinder, remembered that before RAF personnel arrived, her then-fiancé had stood guard over the wreck with a stalwart, if somewhat hidebound, policeman who had pedalled over from Hope: 'He wouldn't even allow Bill into the fuselage to shelter from the rain.' Despite the policeman's august presence, however, Mrs Heardman recalled, too, that when the second party of RAF recovery men begged a supply of hot drinks up at the site, her fiancé suggested that they bring down the sturdy vacuum flasks he had seen in the fuselage. 'Oh!' he was assured, 'they've long gone.'

But clearly, hot drinks or not, the salvage teams did not do that thorough a job, for in 2010 a considerable amount of debris still remained. Perhaps that was not so surprising, though, for as veteran ranger John Campion 'Campy' Barrows observed in 2007, despite retaining much of its integrity the Halifax had nevertheless shed a considerable amount of debris in the course of a relatively lengthy slide; indeed, in the 1980s researcher Mr Alan Jones found tools dropped by the salvage parties within feet of the trig column, far beyond what is now taken as the terminal point.

Visiting the site

To reach the crash site of Halifax HR727 from the Edale public car park means a 360 metre (1,200 feet), 80 minute climb, mostly on reasonable upland paths. Having crossed the footbridge over the gully beyond the village, the path divides, the ascending route to the right offering the most direct route – yet not the most testing – to the relevant area of the rim path. This path initially zig-zags around The Nab (after which the most strenuous stretch is done!) and then, beyond a plateau (at SK 12450 86980, 1,503 feet), branches right to leave the crag of Ringing Roger to the left as it heads north-eastwards for a lowered section (col) on the skyline.

Having reached the rim path at this col, at SK 12949 87393 (1,808 feet), a minor path leads directly upwards following a well-defined water channel (grough) bound for the trig column. A secondary rim path is crossed after just two minutes, a further eight minutes leading to the Halifax's impact site, in heavily broken ground, off to the right. Despite the proven accuracy of the GPS location, be prepared to search around, for by 2010 the debris had become minimal. The terminal point is located three minutes further on and just 140 yards short of the trig column.

When at the terminal point and ready to return to the rim path, the walker might just as well face south-westish (235°M) – or 255°M from the *impact* point – and carefully brave another six minutes or so of very pathless, soggy and grough-bisected moor to visit the crash site of Wellington X3348 (see above). Again, on reaching the co-ordinates, search about to find the minimal debris pool.

On leaving this site another similar stretch of some five minutes will regain the rim path to the west of Ringing Roger. From here, either The Nab or the Golden Clough paths offer a scenic descent of 80 minutes to the car park, the snaking way down to The Nab being the least demanding.

For those wishing to extend the walk, then holding the height and continuing westwards along the undulating rim path – paved in places – for some 30 minutes will lead to a crag overlooking the crash site of Wellington W5719 (see above). Some 15 minutes beyond that, two descent routes to Edale offer themselves. The first down Grindsbrook Clough, and the second, rather longer, over or around Grindslow Knoll, beyond which lies the crash site of Heyford K6875 (see above) on Broadlee-Bank Tor. From there, the rough track following the drystone wall north-eastwards and downwards will intersect with the well-defined path back into Edale. Before doing so, incidentally, the wallside track passes above the bilberry-covered slope from which the carcass of Cessna G-BFRP was cleared (see the Non-Debris section, page 143).

15

Boeing F-13A (RB-29A Superfortress variant) 44-61999

Higher Shelf Stones, Shelf Moor

SK 09042 94912	612 m
Unit and Station	(44-61999 assigned to) United States Air Force, Strategic Air Command, 311th Air Division, 91st Reconnaissance Group, 16th Photographic Reconnaissance Squadron, detached to RAF Scampton from McGuire Air Force Base, Fort Dix, New Jersey
Date	3 November 1948
Crew	**Flight crew, 324th Squadron:** Captain Landon Peter Tanner, pilot, airplane commander Captain Harry A. Stroud, co-pilot Sergeant Charles R. Wilbanks, navigator Technical Sergeant Ralph W. Fields, flight engineer Staff Sergeant David D. Moore, radio operator Staff Sergeant Gene A. Gartner, radio operator Corporal Clarence M. Franssen, radio operator Corporal George Ingram, Jr, radio operator **401st Motion Picture Unit:** Captain Howard Keel, photographic advisor **Photographic section, 16th Squadron:** Sergeant Donald R. Arbogast, photographer Technical Sergeant Saul R. Banks, photographer Private First Class William M. Burrows, photographer Staff Sergeant Robert I. Doyle, photographer

Boeing F-13A 44-61999

Crashed 3 November 1948

The Shelf Moor heights from the parking lay-by on the Snake Road

During the 1948 Russian blockade of Berlin, the United States Air Force utilised B-29 Superfortresses of their Second Bomb Group to augment the transport air-supply fleets. Among these bomber types, however, they interposed some of the F-13 photographic variant to secretly film Soviet-held territory. On 3 November 1948, one of these F-13s, 44-61999, having returned to its temporary UK base, and just days before its planned homeward passage to the States, was detailed for a delivery flight to the main American depot at Burtonwood, near Liverpool.

At 1015 hours on that day Captain Landon Tanner, later to be described as 'one of the most experienced of B-29 pilots', took off from RAF Scampton. The particular machine he was flying, inherited by his 13-man crew complement for the duration of their European detachment, was a veteran in its own right, and much earlier in its photographic career had been named – with what was to become ironic prescience – *Over Exposed!*

The route forecast for the flight reflected that there would be scattered cloud from 2,000 to 4,000 feet, and an overall visibility of 4 to 6 miles, reducing somewhat in occasional rain showers. In view of this Captain Tanner elected to fly under Visual Flight Rules; essentially, to maintain visual contact with the ground, keeping clear of cloud. The distance from Scampton to Burtonwood is just 86 miles, the track westerly, and at a conservative low-level cruising speed of 220 mph (191 knots) the aircraft should have arrived at Burtonwood some 22 minutes after take-off. Except that, initially inexplicably, it failed to do so.

A group of walkers pauses at the crash site

Concern would have grown only gradually, but when 44-61999 was a full hour overdue a contingency plan came into operation, as a result of which a search aircraft reported burning wreckage on high ground some 31 miles short of Burtonwood.

Had there been survivors it would have been truly propitious that this sighting report was picked up by the RAF's Harpur Hill Mountain Rescue Team which was exercising in the area. As it was, although members of the team reached the crash scene with commendable dispatch, they found no one to aid. Before nightfall they had located eight bodies, but had found no sign of life anywhere in the burning wreckage.

The would-be rescue effort truly got under way at first light the next morning,

but although widening moorland searches were initiated, the bodies of the unaccounted-for crew members were eventually found amidst the debris at the scene.

In general, the findings of American air-accident investigations from that era tend to be more benign than those of their RAF counterparts. Certainly, in this case, the investigators seem to have done their best to avoid deciding upon a condemnatory cause, arguing that there were too few facts to enable them to reach any definite conclusion.

The aircraft, they reported, had been found to be in alignment with, and just 3 miles north of, the direct track to Burtonwood; there had been no witnesses and no recorded emergency transmissions. Beyond this, a crew member's watch had been smashed while reading 1050 hours. Taking this as the time of impact, and calculating from the known take-off time to obtain the planned estimate for Burtonwood – around 1037 hours – they might well have reasoned that the crew had done some sightseeing before entering the cloud belt; a reasonable assumption in view of their imminent departure from the United Kingdom. As it was, the Accident Committee had no remit to speculate, and merely submitted that Captain Tanner and his crew had inexplicably flown into the high moors; that consequently the Committee could make no recommendations.

Higher formations accepted this indecisive finding. Nevertheless, they expressed their concern that this appeared to be the type of accident that occurred when a pilot tried to maintain a Visual Flight Rules operation even when the weather conditions became marginal. An observation which, predictably enough, immediately drew avowals from all the Division's subordinate commanders that *their* pilots were constantly urged to re-file to an Instrument Flight Rules plan whenever such conditions were encountered.

The salvage team reduced the wreckage as far as possible, and yet in early 2010 a bewildering amount remained; indeed, one might have said a monstrous amount. Certainly if anyone wished to make the case that such wreckage should be cleared once and for all, then this site might well be proffered as the prime example. That said, walkers have been observed to linger by the adjacent Higher Shelf Stones trig column without having the slightest inkling that the northerly aspect holds anything but moorland, so well do the groughs conceal the remains.

An unofficial – but seemly – monument was set up by RAF personnel in 1988 which, this walker sincerely hopes, will long see out the cult of poppies and wooden commemorative crosses which in late 2009 seemed set to overwhelm the wreckage. Hearteningly, by early 2010 sense had prevailed, with the tribute reduced to an appropriate thirteen crosses.

Visiting the Site

A convenient starting point for all the Shelf Moor sites is at the intersection of the A57 and the Pennine Way (SK 08800 92900). However, the lay-by here offers only limited space, so it might be necessary to park at Doctor's Gate Culvert, half a mile to the east (SK 09600 92900). The Pennine Way leads north-eastwards for something over a mile to Hern Clough (at SK 09700 94800) from which a grassy track lifts westwards towards the 621 metre trig column, a marvellous vantage point in its own right.

In order to visit the other two Shelf Moor sites, that of Skytrain 42-108982 and Lancaster KB993 (see below), it is recommended that the height is held for some 20 minutes (a half-mile) while utilising any track, or convenient ground, that offers.

For the descent, simple back-tracking to Hern Clough is not to be despised, but a return route full of interest is that which descends Ashton Clough, with its spillage of wreckage from the Skytrain, and then follows the Doctor's Gate track back to the Pennine Way, and up to the starting point.

The tasteful display of thirteen commemorative wooden crosses

14

Douglas C-47 Skytrain (Dakota) 42-108982

East of James's Thorn, Ashton Clough, Shelf Moor

SK 08069 94736	528 m, impact site, east
SK 08045 94736	530 m, impact site, west (see below for other, related sites)
Unit and Station	United States Ninth Army Air Force, 9th Air Force Support Command, 314th Troop Carrier Group, 32nd Troop Carrier Squadron, AAF527 (RAF Leicester East)
Date	24 July 1945
Crew	Five crew, two passengers, all killed First Lieutenant George L. Johnson, United States Army Air Force (USAAF), pilot First Lieutenant Earl W. Burns, USAAF, co-pilot First Lieutenant Beverly W. Izlar, USAAF, navigator Sergeant Francis M. Maloney, USAAF, radio operator Sergeant Theodore R. McCrocklin, USAAF, crew chief Corporal Grover R. Alexander, USAAF, passenger Leading Aircraftman John D. Main, RAF, lift-hitching passenger

From 1936 the American Army Air Corps played an active part in the development of the Douglas airliner that metamorphosed into the DC-3 – the Dakota to the British – spawning in turn the C-53 Skytrooper and the C-47 Skytrain variants. By 1941 the Air Corps had become the Army Air Force and the C-47 Skytrain its standard transport aircraft, which, with a strengthened floor and a wide cargo hatch, proved capable of carrying out a seemingly infinite number of tasks. So it was that when C-47 42-108982 was dispatched from the American base at RAF Leicester East to that at RAF Renfrew (subsequently Glasgow airport) it was carrying not only passengers but a jeep, lashed aft.

Dependable as the C-47 was, however, this one was not to complete its flight, for

One of the impact sites

just 69 miles into its 230-mile route it was flown into the cloud-covered hillside high on Shelf Moor where it disintegrated, much of it burning, leaving no survivors.

The jeep, for its part, burst its lashings and was propelled forwards with such violence that it smashed its way through the passenger seating zone and out of the aircraft, so illustrating the danger of stowing main-cabin freight to the rear of passengers. But lessons so taught are soundly learnt! For the practice later adopted on types with no under-floor freight hold has been to load cargo *forward* of passengers when operating in the passenger-cum-freight role.

Before his departure, the pilot, First Lieutenant George Johnson, had been advised to follow an east-coast routing in view of reported strong winds and low cloud ceilings over the High Peak. Despite this he made the perfectly proper command decision to fly the direct, time-saving track. Only less properly – indeed, fatally for his crew – he then chose a demonstrably unsafe cruising altitude, taking too little account of the height of the en-route terrain, and just as little of either the low cloud or the vertical currents the strong winds would breed.

It says much for the rugged construction of the Dakota that, despite the force of the impact and the subsequent fire, the rear fuselage and tailplane were relatively intact when the aircraft was discovered by an off-duty airman, identified by the *Glossop*

Chronicle as a Sergeant Pidgeon, two days later. The salvage teams disposing of the wreckage, however, spilled much of it down the adjacent Ashton Clough.

In early 2010 there were two distinct sites in the area of impact, each with debris and each still largely devoid of vegetation. In addition, below the fence erected in 2003, an undercarriage member still lay on the shoulder of Ashton Clough. Finally there was the rest of the wreckage trail extending down the Clough to Shelf Brook and Doctor's Gate. Until 2000 or so both engines were to be seen, since when the best-preserved one had either been removed by enthusiasts or buried by floods. In which context, speaking of a major airframe section, long since become seemingly integral with an accumulation of soil and rocks, Mr Mike Brown, of Glossop, an inveterate crash-site visitor throughout his boyhood, asserted, 'When we'd first go there, in 1957 or so, that aft-fuselage section, just before the tail, provided a good shelter from the rain, and would just hold three lads.'

Debris spilled into Ashton Clough

Lower sites as recorded in early 2010, all being subject to flood:
SK 08139 94627 478 m, undercarriage components, grass shoulder of the clough
In Ashton Clough, descending:
SK 08155 94594 446 m, fuselage section
SK 08144 94576 440 m, fuselage member
SK 08130 94577 437 m, panel
SK 08107 94496 424 m, undercarriage component
SK 08075 94434 390 m, reduction gear
SK 08067 94317 377 m, engine
SK 08057 94287 355 m, panel
SK 08051 94254 345 m, fragment

In late 2002 a memorial to the co-pilot, First Lieutenant Earl Burns, appeared at the site, placed there by members of his family following research by aviation enthusiasts. It clearly annoyed some walker's sensibilities, for by the following July it had been smashed.

Visiting the Site

The Skytrain site is most easily visited from the lay-by at the summit of the Snake Pass where the Pennine Way crosses the A57 (SK 08800 92900). This lay-by offers only limited parking, so it might be necessary to drive down to Doctor's Gate Culvert, about half a mile to the east (SK 09600 92900).

The access route follows the Pennine Way north-eastwards to Hern Clough (at SK 09700 94800) from which a grassy track climbs off westwards towards the summit. As this is approached it would be a sad walker who did not want to enjoy the view from the rocks bearing the 621 metre trig column. Having looked one's fill, however, and having taken in the enormity of the B-29 site (see above), it is advisable to hold the height and walk westwards for half a mile (some 20 minutes), choosing any suitable track or open ground to pass behind Lower Shelf Stones.

After visiting the double impact site of the Skytrain, the Lancaster debris pool on James's Thorn (see below) is only a five minute walk away.

The most interesting descent from the Skytrain sites is that down Ashton Clough. This is where much of the wreckage was spilled in order to obviate its distracting future air-rescue searches. Before entering the Clough, however, a grassy descent directly below the main sites will lead to an undercarriage member, although the convex slope keeps the item hidden for some while.

In fact, the clough forks at the level of this component. When ascending the clough, either take the left fork or, more rewardingly, emerge onto the grass and bracken slope at this point, see the undercarriage member, then work directly upslope to reach the dual impact sites. The descent of the clough will be taken steadily, of course, the

Crashed 24 July 1945

Debris on the slope below the impact sites

irregularly-spaced surviving fragments, some of which have taken semi-permanent station, adding an extra dimension of interest to what is, in its own right, a very attractive feature. The clough does, however, include two waterfalls each of which needs due care in by-passing.

Emerging from the clough, the climb up to the Doctor's Gate path is made in the vicinity of a tarn. It is then a case of right turn for Glossop, or left turn for the summit of the Snake Pass, alternatively, for the extra half a mile to Doctor's Gate Culvert.

All three Shelf Moor sites are popularly visited from Glossop, often from the long-serving opportunist turning-circle parking place at SK 04580 94830. The route leads eastwards along Shepley Street to Mossy Lea Farm to become the Doctor's Gate path. Although the slopes can be scaled long before reaching Ashton Clough (at SK 07980 94080) one advantage of ascending this picturesque water channel is that pausing periodically to inspect bits of debris covers up the need to stop for breath. The hill-breasting paths, with their spectacular panoramic views, can then be left for the descent.

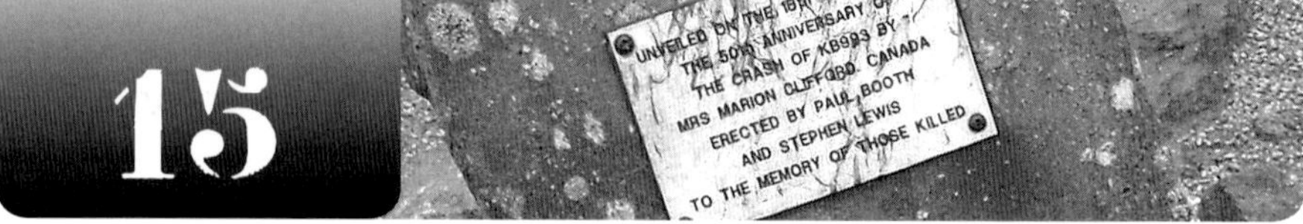

Avro Lancaster BMk.10 KB993

James's Thorn, Shelf Moor

SK 07928 94781	542 m
Unit and Station	No. 408 Squadron, Royal Canadian Air Force (RCAF), RAF Linton-on-Ouse, No. 6 Group, Bomber Command
Date	18 May 1945
Crew	Six, all killed Flying Officer Anthony Arthur Clifford, RCAF, pilot Flying Officer David Fehrman, RCAF, bomb aimer Pilot Officer Kenneth McIver, RCAF, flight engineer Warrant Officer Michael Cecil Cameron, RCAF, wireless operator Flight Sergeant Clarence Halvorson, RCAF, air gunner Flight Sergeant Leslie Claude Hellerson, RCAF, air gunner

Starting in 1941, Lancaster production continued throughout the Second World War with the Avro Company making full use of the capacity of other manufacturers and sub-contractors. Despite this, demand was so high that in 1942 the Victory Aircraft Company was set to producing Packard-built, Merlin-engined Lancasters in Canada, the first of 430 being received in August 1943. Having been flown over the Atlantic, these aircraft, designated Lancaster BMk.10s, were then operationally equipped and allocated to their respective squadrons. One BMk.10, however, Lancaster KB993, was to be tragically lost, not on operations, but on a local detail; and lost, moreover, ten days after the cessation of hostilities in the European Theatre.

On the night of 18 May 1945, Flying Officer Anthony Clifford, of the Royal Canadian Air Force, was tasked to carry out a detail of circuits and local-area flying from RAF Linton-on-Ouse, near York. The squadron had just been given notice of its return to Canada so this detail would have been seen as a chance to relax and to enjoy flying in a way that had scarcely been possible since any of those on board had embarked upon aircrew training. It would also afford them a chance to see what the newly-emergent Britain really looked like at night, although the blackout restrictions had, in fact, been gradually relaxing over the past twelve months.

Because the detail was to involve only local flying, albeit at night, Flying Officer Clifford clearly did not feel it necessary to take his navigator along, for he got airborne with just five of his regular crew. Just the same, as air gunners were hardly required on a peacetime circuit detail, their presence gives credibility to the notion that the forthcoming sortie was regarded as a well-deserved pleasure flight.

Once airborne Flying Officer Clifford soon cleared the circuit, and it might be said, proceeded to stretch the meaning of 'local area' a little too much, for when next seen his aircraft was circling Glossop, 46 miles to the south-west of its Yorkshire base.

It was later established that shortly after this sighting, and while climbing in cloud, the Lancaster was flown into the shoulder of James's Thorn, striking at 1,800 feet

The shrinking debris pool

and bursting into an inferno of flames, a combination which killed most of the crew instantly, with a sole occupant, very seriously injured, surviving only briefly.

One might well wonder that Flying Officer Clifford should steer anywhere near James's Thorn, with its elevation of 1,900 feet above sea level, at night and at such a low altitude. But if, as seems likely, he had become uncertain of his position – being more confused, perhaps, than aided by the unaccustomed, horizon-filling vastness of Manchester's lights – he would have been able to fix himself by craning down from left-hand orbits over Glossop. After which, being more accustomed to having a navigator at his shoulder, he would have had to estimate a line on the map north-east to Linton-on-Ouse. Yet if such a course is actually plotted, allowing for the aircraft to have rolled out from overhead Glossop, then James's Thorn does indeed lie on the track to Linton. So it must have seemed to Flying Officer Clifford that he had things well under control once more as he stopped his turn and climbed away north-eastwards. As indeed he might have had, if only he had searched out the contours as accurately as he had the track, and realised just how little distance he had in hand to climb to any sort of safety height at all.

Neither the inquiry nor senior authority were able to condone Flying Officer Clifford's action in taking his aircraft so far from base while on a specifically 'local-area' clearance. Where the surviving members of the squadron were concerned, and particularly those of Flying Officer Clifford's regular crew who had not been on board, the tragedy must have sadly dampened the anticipation of their return to Canada. Yet, inexorably, the move went ahead on schedule, just a matter of days after the remains of those unfortunates of the squadron had been laboriously stretchered from the moors.

By 2010 the spine of James's Thorn, high above Doctor's Gate, had become a relatively popular footpath; which made it surprising that quite so much debris remained. Especially as in 1995 a memorial column was air-lifted into place by a Sea King helicopter in an operation organised by two enthusiasts, identified by the *Glossop Chronicle* as Mr Paul Brook and Mr Steve Lewis, both described as gardeners for the New Mills Council. Then, on 22 May 1995, a month or so later, a ceremony was held on the site with some participants being helicoptered in. Foremost among them was Mrs Marion Clifford, the 91 year old mother of Flying Officer Clifford, she and other relatives being picturesquely supported by pipes and drums from the Clan Urquhart Highlanders. The monument, movingly, pays tribute not only to the Lancaster crew but also to other Allied airmen who died close by, for the combined plaque indicates too the location of the American cargo aircraft, Skytrain 42-108982 (see above).

Visiting the Site

The site is most easily visited from the lay-by at the summit of the Snake Pass where the Pennine Way crosses the A57 (SK 08800 92900). This starting point offers only limited parking, so it might be necessary to drive down to Doctor's Gate Culvert, about half a mile to the east (SK 09600 92900).

Crashed 18 May 1945

This access route to the grandly imposing Shelf Moor Range – surely the least laborious way for a Sunday walking party to get from road to rocky summit! – follows the virtual highway of the Pennine Way north-eastwards to Hern Clough (at SK 09700 94800) from which a grassy track leads off to climb westwards towards the 621 metre trig column. No walker would want to miss the view afforded from the summit rocks but having taken this in, then the height should be held – not least to enjoy to fullest advantage the continuing picture-postcard vistas from that level – and any convenient track or easy ground followed westwards for some 20 minutes, just over half a mile, to reach the James's Thorn site.

The tasteful monument, a fair pattern for others

It would, of course, be unusual on such a walk to have passed unvisited the sprawl of the B-29 site in the immediate vicinity of the trig column (see above). Indeed, on the approach to the summit it is more than likely that discarded pieces of metal will have drawn attention to it, off to the right.

Another popular approach to all three Shelf Moor sites is from Glossop, a long-serving parking place being the turning circle at SK 04580 94830. This access route leads eastwards along Shepley Street to Mossy Lea Farm to become the Doctor's Gate path. Although there are various tracks up the hillsides, the ascent up Ashton Clough (from the tarn, just below the path, at SK 07980 94080) has to be that of choice. The steep, grassy paths can then be left for the descent.

In fact, the careful descent of Ashton Clough, with its spillage of Skytrain wreckage, is also the most interesting way down (back-tracking to Hern Clough being the main option), offering gloriously confined scenery in contrast to the panoramic exposure experienced while transiting the ridge routes. On issuing from the clough, the way then follows the Doctor's Gate track, climbing either east to the Snake Pass summit starting point, or descending west, to Glossop.

16

De Havilland (Canada) L-20A Beaver 52-6145

Bramah Edge, Torside, east of Glossop

SK 05531 97589	399 m
Unit and Station	United States Air Force, 81st Fighter Bomber Wing, 7519th Air Support Group, Operations Squadron, RAF Sculthorpe
Date	5 December 1956
Crew	Pilot and passenger, both killed First Lieutenant John Rossman Tinklepaugh, USAF, pilot First Lieutenant Guy B. Waller, USAF, passengering pilot

The L-20A Beaver, manufactured by de Havillands of Canada between 1952 and 1960, was used by, amongst others, Britain's Army Air Corps. The United States Air Force, though, operated 200 Beavers as liaison and light-transports, one of which, Beaver 52-6145, was on the strength of the 7519th Air Support Group Operations Squadron when it made its last flight on 5 December 1956.

It was flown on that occasion by First Lieutenant John Tinklepaugh, who had been detailed to transport fighter pilot First Lieutenant Guy B. Waller from Sculthorpe (east of the Wash) to the American base at Burtonwood (Liverpool), to pick up an F-84 jet fighter. A straightforward enough task which merely required First Lieutenant Tinklepaugh to conform to an initial departure routing, then fly 140 miles to the radio beacon at Oldham where he would turn south-westwards for a radar pick-up and descent into Burtonwood.

As significant medium-level cloud cover was forecast, and being aware of the high ground obtaining as far as Oldham, First Lieutenant Tinklepaugh filed an instrument flight plan for an altitude of 4,500 feet. Additionally, using the forecast weather, he calculated that his flight would be slowed by a headwind of 20 knots, resulting in a groundspeed which would bring him overhead Oldham some one hour and 20 minutes after take-off, and over Burtonwood 16 minutes later.

What was to remain unknown to him was that, because of an approaching cold front, the headwind had more than doubled in strength, which meant that it would now take him an additional 30 minutes to reach Oldham, and an additional six minutes to reach Burtonwood. Such unallowed-for extensions to his expected journey times set the scene for the tragedy about to be played out.

And very unusually, a tragedy in which mistakes were not only to be made by the man in the air, but also by the controllers on the ground. Perhaps, therefore, it would be germane to recap on the procedure for the acceptance of an aircraft by air traffic control radars in those far distant days.

For many years, identification of a target aircraft has been made by its on-board transponder equipment automatically answering any interrogating radar with a dedicated code, this code (or 'squawk') being assigned by air traffic control before take-off. Indeed, even back in the 1950s some aircraft would have carried the transponder's predecessor 'Identification Friend or Foe' (IFF) – codenamed 'Parrot', from which the still current 'squawk' procedural words survive – but clearly Beaver 6145 was one of the many which did not.

In those days, accordingly, the pilot would advise radar of his estimated position and heading. Radar would then monitor any targets showing on the screen in that vicinity which seemed to correspond, at the same time instructing the aircraft to turn through 30 degrees for identification. And provided a target on the screen complied, and turned back again when so directed, positive identification was assumed and radar control was established, such identification being reinforced as the target continued to reflect radar's acknowledged instructions.

Regarding the radio-navigational aids fitted to this Beaver, it is not known whether First Lieutenant Tinklepaugh had the benefit of a (serviceable) radio compass to tell him when he was overhead the Oldham beacon, although it can be deduced from subsequent events that this facility was lacking. Further, it should be appreciated that in the fifties even a communications radio was a relatively low priority.

Reflecting upon what is known of the conduct of the flight itself, First Lieutenant Tinklepaugh's departure clearance required him to make a sizeable dog-leg after take-off in order to avoid the Wash danger areas, then to call when he was overhead RAF Marham. In the event he did not call, as a result of which, it was later held, he could not be given the updated en-route weather, including the significantly increased upper-wind speeds. However, it is a matter of record that he subsequently obtained a position line from a directional facility abeam his track by means of which, as evidenced by his later estimates, he realised to some extent at least that he was falling behind his planned timing. Unfortunately, his calculations failed to show him just how far behind.

The flight path from then on becomes complex, but by using the air traffic record to back-plot from the crash site, a reasonably clear picture emerges.

At 1213 hours, after unknowingly combating a greatly increased headwind for an hour and 26 minutes, First Lieutenant Tinklepaugh reported that he was overhead

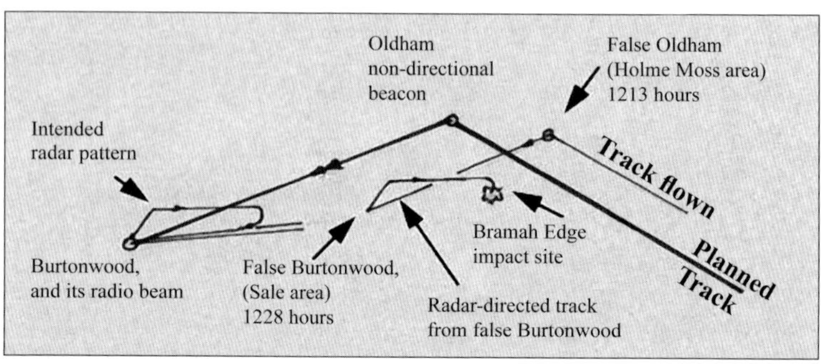

The track flown by Beaver 52-6145

the Oldham beacon. As previously said, what navigational means he employed to determine this is not known, but on the strength of his call Air Traffic cleared him down to 3,500 feet towards Burtonwood. In fact, he would have been nearer Holme, reasonably on track from Marham, but still some 8 miles short of the Oldham facility.

As directed, he then contacted Burtonwood, advising that he was now at 3,500 feet and inbound to Burtonwood on a south-westerly heading. Subsequent to this Burtonwood radar, tracking likely-looking targets approaching from the Oldham beacon area, twice ordered him to make identifying turns. On each occasion, however, as the target on the screen failed to respond, First Lieutenant Tinklepaugh was directed to resume his own navigation for Burtonwood, with the additional anticipatory instruction that, having reached it, he should turn onto a north-easterly heading.

No positive radar identification was made, but at 1228 hours First Lieutenant Tinklepaugh reported that he was overhead Burtonwood and turning north-east; in fact, he was most probably in the Sale area. But on the radar screen a target was seen to be moving to the north-east from Burtonwood, and when, a minute or so later, First Lieutenant Tinklepaugh acknowledged (and obeyed!) the instruction to turn easterly – onto what was intended to be a long, downwind leg – this target did the same. Later investigation would prove this target to have been an aircraft coincidentally flying a visual downwind leg below the 1,600 foot cloud base, and on the Burtonwood tower frequency, not on the radar frequency. The reality was that First Lieutenant Tinklepaugh had been steered, initially north-eastwards towards Salford, then eastwards to the Crowden area.

Some minutes later, radar, now convinced that this was a positive identification, ordered a wide right turn intended to bring Beaver 6145 back to line up with the Burtonwood runway, clearing it to descend in the turn to 1,500 feet. Only although First Lieutenant Tinklepaugh immediately acknowledged that he was conforming, the

Crashed 5 December 1956

A contemporary photograph of the crash scene

target on the screen did not turn! Instead it continued to track eastwards.

Initially just a little puzzled, radar then made a querying call to which First Lieutenant Tinklepaugh promptly responded, confirming that he had indeed made the right turn and was well into the descent to 1,500 feet.

More hurriedly now, radar advised him to hold 2,000 feet and re-home to Burtonwood. But the only response to this instruction was First Lieutenant Tinklepaugh's startled, and abruptly truncated exclamation, '*What – !*' Then silence.

And in that silence the misidentified target on the screen continued on its way, fading from sight as it reached ten miles east of Burtonwood. While a full *thirty* miles east of Burtonwood, Beaver 6145 was now a burning mass amidst the boulders 1,300 feet up on Bramah Edge, and both its occupants were dead.

Mulling over the failure of the pilot, albeit with just 550 hours' experience, to fully utilise his in-flight track-check to more accurately revise his groundspeed, it is noted that his passenger, in the right-hand seat alongside him, was a jet-fighter pilot: an ebullient breed, and prone to indulge in raillery when passengering aboard a slow transport. A formula, if not for disaster, then certainly for distraction. But after all, where was the need for undue concentration – 4,500 feet was a safe height. And once in the vicinity of the Oldham beacon, Burtonwood radar would safely gather them in…

The debris pool, looking down onto Torside and the B6105

Visiting the Site

The most convenient route to the site starts from Torside, on the B6105, to the east of Glossop, where there is adequate parking at the Torside public car park. A footpath from the rear of the car park leads up to the Longdendale Trail which is followed westwards for a mile or so until it intersects with the Pennine Way below Reaps Farm. Alternatively, very limited lay-by parking on the B6105 can be found at this intersection (SK 05725 98044).

From here the route snake-climbs past Reaps Farm and up the Pennine Way to SK 06330 97630 where a drystone wall leads off to the west. An intermittent rim track, often on the line of the broken wall, then presents little difficulty during the near half-mile it takes to reach a position abeam the dam separating the Torside and Rhodeswood reservoirs.

In early 2010 a small cache of debris remained at the crash site but this is one of those sites where, given the most trusty of references, a visual cast-about will probably be called for, in this case because of the proliferation of overhanging boulders. Having edged over the rim at this point, the boulder-spill also demands care. The view afforded over the valley, though, is superb.

17

Bristol Blenheim Mk.1 L1476

Sykes Moor, Torside Clough

SK 08295 97035	481 m
Unit and Station	No. 64 Squadron, No. 12 Group, Fighter Command, RAF Church Fenton, York
Date	30 January 1939
Crew	Both killed Pilot Officer Stanley John Daly Robinson, pilot Acting Pilot Officer Jack Elliott Thomas, passengering pilot

First flown in early 1935, the Bristol Type 142 was developed as the Blenheim bomber, the RAF receiving the type in March 1937. Nearly two years later, however, on 30 January 1939, with the type well established in the Service, Blenheim Mk.1 L1476 disappeared with its two South African occupants, Pilot Officer Stanley Robinson and Acting Pilot Officer Jack Thomas, both of whom had been engaged in a local-area familiarisation flight from RAF Church Fenton, to the south-west of York: 'Local flying and sector knowledge', as the RAF accident summary specified.

Although a search was mounted it was redirected at an early stage, and eventually called off, following a report – subsequently found to be false – that an aircraft had crashed into the sea. Indeed, the missing Blenheim was not discovered until thirteen days later, on Sunday, 12 February 1939, when a chance walker, identified by the *Glossop Chronicle* as Mr Richard Robert Bridge, came upon its wreckage.

The senior officer appointed to investigate the crash, Squadron Leader Heber-Percy from RAF Church Fenton, found a sizeable crater suggestive of a high-velocity impact, with debris thrown outwards over 'a couple of acres' by the exploding fuel tanks and with both engines detached, one being a hundred yards from the crater. The *Glossop Chronicle*'s account catalogued the condition of the bodies in what some must have felt was unnecessarily lurid detail. In essence, however (and as the Glossop Coroner's Court was told), Pilot Officer Robinson, together with his streamed – but undeployed – parachute, was found 600 yards from 'the main wreck', while Acting Pilot Officer Thomas lay within 200 yards of his designated captain, in a direct line from

The contemporary scene

the tail (showing that Pilot Officer Robinson had ordered an abandonment and that both men had jumped; but at far too low a level).

Variously questioned by the coroner, Mr G.H. Wilson, and also by Police Sergeant Clarke of New Mills, Squadron Leader Heber-Percy gave it as his opinion that it was not unusual after a high-speed impact for the fuel tanks to have exploded without catching fire, the explosion not even staining the aircraft's fabric, but merely scattering the debris.

Significant evidence was given by a Mr Thomas Ingledew Hardcastle, of Hillside Cottage, Glossop, who told the court how, with the moorland tops in mist, 'the sound of an aeroplane with the engines making a loud noise' had caused him to run up a field in a vain attempt to see what was happening. He described how the engine noise had then 'suddenly stopped'.

Despite all of which Squadron Leader Heber-Percy was obliged to advise the court that although the indications were of a high-velocity impact with engines running, no definite cause for the loss of control leading to the crash had so far been determined. Nor does the RAF accident card summary, raised subsequent to the inquest hearing, record any cause. It might be ventured, nonetheless, that either disorientation in cloud or failure to contain an engine-out condition – real, or purposefully entered into as a drill (or indeed, both in conjunction) – were likely causes; particularly as Pilot Officer

The crash site in 2010, with the monument, looking towards Torside

Robinson, the aircraft captain, could claim only ten hours on multi-engined types. True, he had amassed a reasonable enough 363 hours in his two years as a pilot, but (having trained in South Africa) few of these would have been gained in the temperate-zone weather conditions of a British winter.

One facet of this accident takes us back, as to a bygone age. For when the chance walker, Mr Bridge, descended the moors to Reaps Farm, above Torside Bridge, it was to find that the farmer, Mr Bert Crossland, had no telephone. The bleak news, therefore, had to be passed via 'a railway telephonic message' from the nearby Torside railway signal-box to the railway controller at Manchester, who was able to alert the Glossop Borough police so that, in his turn, the Derbyshire County Police Force officer responsible for the moorland area of Charlesworth parish, a Constable Clark, was detailed to attend.

On the day following the discovery, once the site had been thoroughly scrutinised by his team, Squadron Leader Heber-Percy gave permission for the bodies to be stretchered down to Reaps Farm and taken to the mortuary at Glossop and thence to RAF Church Fenton. By the third day, with the investigators' deliberations complete, and with the salvage team having camped on site, most of the wreckage had already been buried in a gully adjacent to the impact point. Long before 2010, however, much of this had re-surfaced. Just the same, the moor is relatively flat in the area, deeply heather-covered and fretted by many watercourses, so that despite the abundance of debris it may well be necessary to scout around from the reference in order to actually locate the gully and the memorial pillar.

Visiting the Site

There is adequate parking at the Torside public car park, off the B6105. Alternatively there is very limited lay-by parking where the Pennine Way crosses the road and the Longdendale Trail, below Reaps Farm (at SK 05800 98000).

Having climbed past Reaps Farm, the route follows the Pennine Way up Torside Clough for just under two miles to an unnamed water channel at SK 07700 96800. The crash site, with its (unofficial but tasteful) monument, will be found half a mile (ten to fifteen minutes) up this grough.

For those going on to Bleaklow Head an undemanding half-mile heather-tramp southwards will take them back to the Pennine Way.

For what it is worth, the crash site of Wellington R1011 (see below) on Birchen Bank Moss – the eastern extension of this moor – is 1.7 miles distant as crow and GPS has it, on 060°M. This tortuous way lies over substantial areas of reclaimed heather moor and fascinating heather bog. The odd hare may be encountered, and a startling grouse or two, but the drainage is unrelentingly against the line of progress, with not a path to be seen. Certainly, not a trek to be recommended… Should anyone really wish to visit both sites in a single outing, far better to follow the Pennine Way to Bleaklow Head, then the path down Near Black Clough until abeam the Wellington site.

18

Vickers Armstrong Wellington Mk.1C R1011

Birchen Bank Moss, Bleaklow

SK 10544 98580	486 m
Unit and Station	No. 28 Operational Training Unit (OTU), RAF Wymeswold, No. 93 Group, Bomber Command
Date	30 January 1943
Crew	Three killed, two injured Flight Lieutenant Anthony Winter Lane, OTU staff pilot, killed Pilot Officer Charles Douglas Brown, bomb aimer, killed Sergeant Raymond Gerard Rouse, OTU staff wireless operator/air gunner, killed Pilot Officer Grisdale, navigator, injured Sergeant Miller, wireless operator/air gunner, injured

Although the Wellington was withdrawn from bomber operations in October 1943 it was well suited to the needs of the Operational Training Units (OTUs) whose function was to meld the various aircrew specialities into first-line bomber crews. During the course those trainee crew members not essential to the task in hand would be replaced by specialist OTU instructors who would 'screen', or oversee, the crew member carrying out the relevant exercise; indeed, at times, when a shortfall had occurred, OTU staff would actually fill key roles. This happened on 29 January 1943, when the final night cross-country flights of the course were dispatched from No. 28 OTU at RAF Wymeswold, near Loughborough. Among the stream that night was Wellington R1011, captained for the occasion by OTU staff-pilot Flight Lieutenant Anthony Lane.

The round-trip routing of this final trip of the course, designed to introduce the trainee crew members to the sortie lengths they would face when they raided deep into Germany, had taken them to the north of Scotland. At about 0145 hours, however, having been airborne for some six hours and 20 minutes, Flight Lieutenant Lane and

The debris pool, looking towards the A628

his part pupil, part OTU-staff crew were just an estimated 20 minutes from base. With the task so nearly completed, it seems that they were desirous of finishing in style: as close as possible to ETA and exactly on their planned track. For despite the heavy cloud through which they were flying, and which had prevented them from positively determining their position for some time, the decision was taken to descend and obtain a visual fix.

When it came to making this always weighty decision Flight Lieutenant Lane, with very nearly 1,500 flying hours, had vastly more experience to draw upon than most Service pilots of his day. Further, if he was relatively new to the Wellington with just 66 hours on the type, he was not short of navigational assistance for in addition to the trainee navigator, who had already proved his capability over the last six hours, he was carrying among his somewhat truncated crew another OTU instructor, Sergeant Raymond Rouse, an experienced wireless operator/air gunner.

Crashed 30 January 1943

Although Sergeant Rouse was a qualified gunnery leader, his function on this flight was to tone up the wireless-operating performance of Sergeant Miller, the trainee wireless operator/air gunner. Patently this task had been satisfactorily accomplished, for when the decision was made to begin the descent Sergeant Rouse took station behind his OTU pilot colleague the better to assist with the lookout.

Flight Lieutenant Lane was well served, then. Yet despite all his experience, despite the services of two communications specialists on his crew who might have furnished a confirmatory W/T fix (wireless telegraphy – morse code), and despite the availability of a whole array of external get-you-home aids, just before six and a half hours into the sortie he eased back the throttles, and lowered the nose of the aircraft into a blind descent through cloud…

What happened then is succinctly stated on the official accident report: 'Screen pilot tried to pinpoint himself when off track at the end of a cross-country flight. Struck high ground at 1,700 feet in low cloud and rain.'

All very dry and unemotional. But then the court of inquiry was recording an occurrence that had become only too commonplace. As the Air Officer Commanding's curt recommendation makes clear: 'N.F.A.', he penned. No further action.

After all, what more preventative action was there left to take? For descending through cloud without having obtained a positive fix had already killed only too many crews; destroyed only too many aeroplanes. Yet there were so many alternative courses of action and an abundance of flying instructions against the practice, the latter all listed in the *Air Navigation* manual of the day, the *AP1234*, which, having spelled them out then thundered, 'Above all do not descend below cloud unless absolutely sure there is no high ground nearby', in bold lettering in the 1941 original.

Even as he pushed the report aside the Air Officer Commanding would have been only too well aware that, just 30 minutes after R1011 fell into the blind-descent trap, so Wellington R1538, another aircraft of the same stream but almost 35 miles adrift and half an hour over estimate, had done the same at Cellarhead, near Stoke. Again, 'N.F.A.'

The reports completed, it would have been time for the senior officer to allow his thoughts to turn to the rest of that night's stream. To wonder, for example, how many of those who had landed back at Wymeswold uneventfully and were now operations ready had taken similar chances. Only got away with it – on this occasion.

Visiting the Site

The nose-to-tail lorries on the vastly overworked A628 Sheffield-Manchester trunk road can be clearly seen from the Wellington site. Despite its proximity to the road, however, the site seems strangely remote, divorced in its heathery wilderness from the seemingly pastoral lower slopes of the Pikenaze and Withens Moors across the road, softened by distance though these may be. Perhaps this remoteness explains why a tasteful commemorative tile has survived undamaged.

Parking is best found along the slip road which leaves the A627 at SK 118400 99600. This leads down to a riverside path which after 500 yards allows access to the picnic-worthy lower reaches of Near Black Clough. A shooters' path on the right, however, at SK 11700 99360, leads upwards from this idyll onto Birchen Bank Moss, the path representing, in effect, a virtual highway to Bleaklow Head.

Getting from the footpath to the Wellington does require cutting across heather moorland for at least 500 yards. And moorland, at that, where the many drainage channels are steadfastly running across the necessary line of advance. The best that can be said is that the groughs in this vicinity are not all that deep. Nevertheless, an exercise best made at the start of a foray, rather than when tired and out of puff at day's end.

If the plan is to carry on to visit the other local sites, across Near Black Clough on Bleaklow (the Defiant and the Botha, see below), then it is best to track the shortest distance back over the rough, then make time along the path to a convenient spot for crossing onto Bleaklow. Just the same, covering the scant few hundred yards to get back to the path may well take of the order of 20 minutes.

The commemorative tile at the crash site

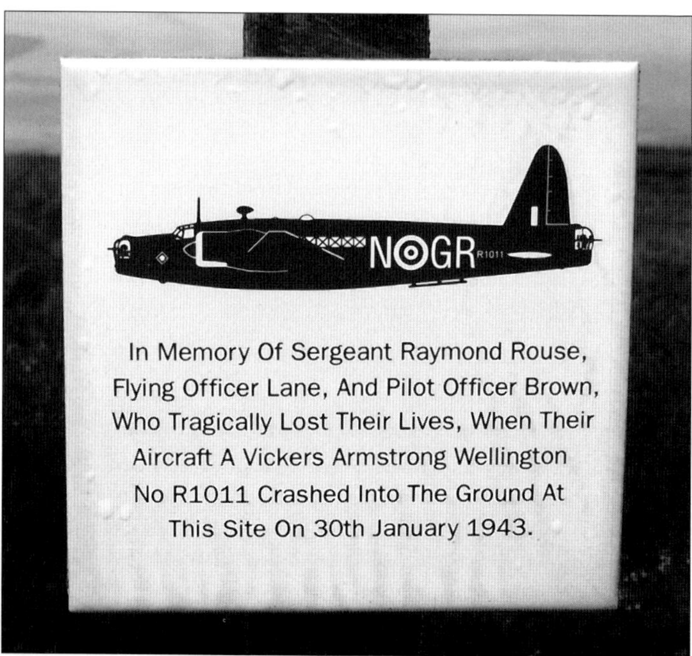

19

Blackburn Botha Mk.1 W5103

Round Hill, Black Moss, Bleaklow

SK 11073 97536	559 m
Unit and Station	No. 7 Ferry Pilots Pool, RAF Sherburn-in-Elmet, Air Transport Auxiliary
Date	10 December 1941
Crew	Pilot, solo, killed First Officer Thomas William Rogers, Air Transport Auxiliary

Throughout the history of aviation there have been British designs which were really not quite the thing, and the Blackburn Botha must surely be reckoned among the foremost of these. Over 400 were ordered in 1936 to provide Coastal Command with a modern torpedo-bomber/reconnaissance machine. However, because priority in engines was given elsewhere, the Botha introduced into RAF service in December 1939 proved unacceptably underpowered for an operational machine. Yet an under-rated engine was only one of the Botha's problems, for it was also prone to certain control complications in manoeuvre and was soon relegated to the training role. Even there, and despite a more powerful engine being fitted, it was found to be unsatisfactory and eventually ended up as a target tug.

On 10 December 1941, First Officer Thomas Rogers, a ferry pilot with the Air Transport Auxiliary (effectively, the 1938 Civil Air Guard, as reconstituted in 1939), was charged with delivering a new Botha from the Blackburn factory at Sherburn-in-Elmet, just east of Leeds, to No. 48 Maintenance Unit at RAF Hawarden, near Chester. It was an 80-mile flight and should have taken the Botha something like 30 minutes. However, the machine never did arrive, and when its wreckage was discovered by a sheep farmer the next day it was 5 miles south of its planned track, although less than halfway along the route.

The machine was found at 1,800 feet above sea level on Bleaklow's Round Hill, inverted, and with the pilot's body suspended in his harness. Yet for all the Botha's high accident rate, and after a painstaking examination, the investigators were unable to fault the aircraft itself. What they could be clear about was that First Officer Rogers had definitely crashed, as opposed to being compelled to make a forced-landing.

Therefore, as Air Transport Auxiliary pilots were generally constrained to maintain visual ground contact, and were often restricted to a maximum height, the conclusion had to be that First Officer Rogers had encountered cloud on the hills and, keen on getting the job done, had felt that by edging aside somewhat he could ease between the mist and the moor; as it turned out, a fatally erroneous feeling.

There was still a substantial amount of debris to be seen in 2010 despite the fact that in 1968 the engines were removed from the moor as a project for officer cadets from the RAF Technical College at Henlow.

Visiting the Site

The Botha crash site is relatively remote for the area but is closer to the A628 (Manchester-Sheffield road) than to the A57 (Snake Pass road) at Doctor's Gate. To approach from the A628 the best parking is found down the slip road at SK 11840 99500, opposite Ironbower Moss Rocks, and easier seen on the ground (or on Google Earth!) than on the map.

This paved road gives way to a track which follows the River Etherow south-eastwards to the mouth of Near Black Clough. Turning into the clough reveals an area well worth lingering in, but for those with business on the moors a shooters' path zig-zags off and upwards at SK 11670 99350, so beginning a south-westerly track – among the finest of moorland paths! – which sits on the shoulder of Near Black Clough and in just over two miles leads to Bleaklow Head.

Long before that, however, at SK 10830 98003, It is necessary to cross onto

RAF Henlow engineering cadets remove one of the Botha's engines

Crashed 10 December 1941

The southernmost wreckage pool

Bleaklow. This calls for a descent to the stream bed of Black Clough followed by the soggy, mossy clamber up the far bank to achieve the track which ascends the unnamed clough opposite. This runs southerly for just under half a mile before opening out, first to broken ground and then to the upper moorland in the area where the debris lies.

An alternative route is that from the Doctor's Gate lay-by on the Snake Pass road. Here, the Pennine Way is followed to Bleaklow Head, at which point two routes offer. The first leaves the Pennine Way to descend on the shoulder of Near Black Clough, when the same unnamed clough will again afford access to the Botha site. The second route – marked by 'stakes', the map promises – runs east to Bleaklow Hill. From there a largely trackless route leads northwards, first past Near Bleaklow Stones and the debris of Defiant N3378 (see below), then, a half-mile later, over very broken, very boggy terrain, to the Botha.

It should be advised that this whole Bleaklow area is worthy of caution. Not only can it be very marshy, but when the mist is down it is only too easy to get disorientated. Even the 'staked path' needs care, for there are few stakes, while comforting boot-tracks tend to mysteriously disappear. A map, compass, and a GPS are an even greater comfort than boot tracks, and when all three conspire against the senses, a cool head is best of all.

20

Boulton Paul Defiant N3378

Near Bleaklow Stones, Bleaklow

SK 10610 96957	574 m, impact point
SK 10649 96948	594 m, terminal point
SK 10624 97052	595 m, gully
Unit and Station	No. 255 Squadron, RAF Hibaldstow, No. 12 Group, Fighter Command
Date	29 August 1941
Crew	Pilot and passenger, both killed Pilot Officer James Craig, pilot Aircraftman Second Class George Daniel Hempstead, passenger

After its, at best, patchy career as a day-fighter, the Defiant fared much better as a night-fighter, its Airborne Interception radar (AI) enabling it to intercept and shoot down enemy aircraft as no other contemporary night-fighter was able to. Losses occurred, nonetheless, as when Defiant N3378 failed to arrive after a seemingly straightforward non-operational positioning trip on 29 August 1941.

N3378 was the machine Pilot Officer James Craig had been detailed to take from RAF Turnhouse, Edinburgh, on the 200 mile – say 45 minute – flight to his base at RAF Hibaldstow, which was effectively, if not at that time in fact, the satellite of busy RAF Kirton in Lindsey, both south-west of Brigg. Nothing appeared untoward, indeed 36 minutes after taking off with an airman passenger (Aircraftman Second Class George Hempstead, identified by the *Glossop Chronicle* as Bolton's golf professional in civil life), Pilot Officer Craig transmitted the routine advisory that he was in contact with RAF Kirton in Lindsey and inbound to his base; Hibaldstow being located abeam Kirton and just 6 miles nearer Brigg. After that, however, there was no further contact and no trace of N3378 or its two occupants until a month later, when its wreckage was discovered at Near Bleaklow Stones, a remote moorland location just below Bleaklow Head, and 46 miles off its intended track.

The preliminary court of inquiry was held while the aircraft was still listed as 'missing',

Crashed 29 August 1941

Pilot Officer Craig's puzzled commanding officer writing, 'Flying experience of pilot makes disappearance difficult to understand'. And clearly, in the month before it was found, the disappearance of Pilot Officer Craig's aircraft was indeed something of a mystery, for the visibility on his planned route had been good, although the cloud was known to have moved in over the high moorlands to the south-west. It should be stressed, however, that for all Pilot Officer Craig's relative experience – 630 hours' total flying time and 100 on type – the true source of puzzlement was not that he had crashed, but that his aircraft had not been found. Certainly, when the court was re-convened, after the missing airmen had been located and the wreckage inspected, any difficulty in comprehension was dispelled. This is shown by the finding added to the summary report to record that Pilot Officer Craig had attempted to blindly penetrate low cloud – in direct contravention of the controlling No. 12 Group's order forbidding such a risky practice – and had flown into a hillside.

Puzzlement, in fact, was replaced by poignancy, the discovery of the wrecked machine having revealed that, although seriously injured, the occupants had survived the actual crash, for they were found seated outside the aircraft, beneath the cockpit; only both had died long before succour arrived.

All this notwithstanding, over the years, and with a fine disregard, at the very least, for the radio traffic logged between Pilot Officer Craig and various ground stations,

The main impact site

air crash enthusiasts would speculate upon more sensational causes than that recorded by the court of enquiry.

One of these was that Pilot Officer Craig had indulged a whim to overfly a house holding members of his family in Wakefield; except that Wakefield would have only have taken him some 20 miles off track, while the Bleaklow crash site is more like 40 miles displaced, as well as being 15 miles further down the planned-track distance than Wakefield. Nor is there any record of the aircraft having been seen over Wakefield, or the house having been buzzed.

An alternative hare started by the mystery-at-any-cost lobby, was that the Defiant had been misidentified and shot down by Spitfires 'near Bishop Auckland': not that any Spitfire incident reports are furnished in support. The enthusiast claim also lends significance, with equal lack of substantiation, to reports that bullet holes were subsequently found in the wrecked Defiant's radiator cowling. But with Pilot Officer Craig's routinely-phrased radio call having been made some 50 miles south of Bishop Auckland one might have thought he would have mentioned, even in passing, being attacked by Spitfires.

Retired schoolteacher Mr Mike Brown, of Glossop, a veteran crash-site hunter, was dismissive of this shooting-down aspect of the Defiant crash. In the course of mutually regretting the scourge of moorland myths which had arisen in recent years, he reasoned, 'The whole moorland where the Defiant came down was used for battle training; you've found 0.303 inch calibre cartridges, both unused and spent, as I have. And what better target to loose off at in such featureless terrain than a crashed aircraft?'

In a bid to course this particular 'mystery' hare, at least, a corrective note might be constructively applied to its most influential perpetuator, *Dark Peak Aircraft Wrecks 1,* p.31 (see bibliography). In all good faith the Collier/Wilkinson coverage, reprising the 1941 event from the late seventies, records that '*it was claimed* [of an army unit] *that they had discovered bullet holes in the wreckage and the theory was put forward that the Defiant had been shot down...*' The passage, however, then takes both claim and theory, adds a rumour, and launches out on speculation: as, of course, is its prerogative. On the other hand, when it observes, '*their existence* [the bullet holes] *was denied in an official report of the day – the official comment on the accident stands at: "Aircraft flew into a hillside in low cloud"*', this, and particularly the expositive dash, is quite mischievously misleading. For the RAF summary crash report, the document in question, denies nothing, makes no official

Debris in a nearby gully

'comment' but, in accordance with its function, baldly states the finding of the court of inquiry, the exact quote being, 'Flew into hill side in low cloud (Group ban)'. The corollary of the official finding being that when the RAF investigators inspected the site there were no bullet holes in evidence; nothing whatsoever, therefore, to deny.

What remains a true mystery is that, with the multitude of causes that would take a single-crew aircraft off track, anyone should bother subscribing to fanciful embroideries. What is clear, from his last radio call, is that Pilot Officer Craig believed that he was nearing his destination; which means that, shortly after making the call, he would have begun a powered descent, expecting to break cloud in the vicinity of Hibaldstow. For the crash site is indeed just 20 or so miles short of the planned along-track distance to Hibaldstow, just right for commencing such a descent had it only *been* Hibaldstow – at 33 feet above sea level – on Pilot Officer Craig's nose and not a Bleaklow moorland elevated to some 1,900 feet above sea level: the delusory height his altimeter would have been showing him when he struck. Truly, with such facts in evidence, only the most dedicated sensationalist would find the need to look any further.

As late as March 2010, a considerable amount of debris was gathered in a pool at the terminal point, additionally there was wreckage at the impact point and many scattered fragments in adjoining gullies. In the mid-1990s a tasteful tribute was positioned in the shape of a well-conceived commemorative tile, but in 2003 this was maliciously smashed.

Visiting the Site

The first of the two main routes to the Defiant site is from the south, from Doctor's Gate, on the A57. The second is that from the A628, the Glossop-Sheffield trunk road, at Ironbower Moss Rocks.

Considering first the approach from the A57, parking is limited at the summit of the Snake Pass but more space might be found at Doctor's Gate Culvert, a half-mile to the east. The walking route then follows the well trodden and often paved Pennine Way to Bleaklow Head. At this juncture two choices offer themselves. The first is to follow the very sparsely stake-marked path to the east, as far as Bleaklow Hill. At this point a northerly turn leads to Near Bleaklow Stones, a third of a mile distant, with the Defiant's three debris sites lying just a little further on. The other choice is to take the path down Near Black Clough, crossing the stream when abeam the site and striking across country.

The approach from the north, from the A628, gives adequate parking where a slip-road leaves the A628 (SK 11840 99500), opposite Ironbower Moss Rocks. From the parking area, the concrete road leads to a footpath which follows the River Etherow to the mouth of Near Black Clough. This stretch well deserves a visit in its own right, but at SK 11700 99360 a shooters' path zig-zags up to the moor. A fine, moorland path then leads along the shoulder of the clough, approaching the Defiant site after just over a mile. Leaving the path and crossing the brook at about SK 10550 96550 will afford a passage along the groughs (water channels), a far easier way to travel than having to cut across a succession of them.

This is a site, particularly in misty weather, when it may be necessary to search about, looking upon the hags as well as on the flatter ground below in order to find all three lots of debris. And a further note of caution: the area lends itself to disorientation, particularly in mist. Nor is the staked path back to Bleaklow Head all that easy to follow. Careful map and compass work, however, with a mapping GPS for choice, will save the day.

21

Airspeed Consul TF-RPM

Crow Stones Edge, north of Howden Reservoir

SK 17403 96616	483 m
Operator	Formerly Adie Aviation Ltd., Croydon, on sale delivery to Iceland
Date	12 April 1951
Crew	Two crew and one passenger, all killed Captain Pall Magnusson, Icelandic Airlines, pilot Mr Alexander Watson, Morton Aviation, Croydon, wireless operator Mr Johann Rist, passenger

The twin-engined, six-passenger Airspeed Consul was the civilian conversion of the RAF's wartime Oxford trainer. It filled a need after the Second World War with many small operators, particularly emergent companies, welcoming the relatively inexpensive machine. Consul TF-RPM had started life as RAF Oxford Mk.I HN471 before being re-registered as G-AHJY and serving with various small United Kingdon companies. On 12 April 1951, having been sold by Adie Aviation of Croydon, it was being delivered to Iceland when it crashed on Howden Moor, killing all three on board.

Captain Pall Magnusson had planned a visual flight with refuelling stops at both Liverpool and Prestwick en route to Iceland. On reporting for departure at Croydon, however, he learnt that the weather pattern was such that westerly winds of up to 60 knots were driving low clouds before them; conditions which made it unlikely that visual ground contact could be maintained. On the other hand, the direct route for Liverpool, crossing Birmingham and then the Cheshire Plain, would not require overflying high ground of any significance. Regardless of the unpropitious weather, therefore, Captain Magnusson duly departed on a visual flight plan.

What happened after that has to be, in part, a matter of conjecture. For the Consul's wreckage was discovered 40 miles to the right of the direct track and in the very centre of the high moors of the Peak District, moors which lift to nearly 2,000 feet above sea level.

The air accident investigators must have deliberated over the time and distance flown along track. The planned distance to Liverpool was some 185 miles – or about

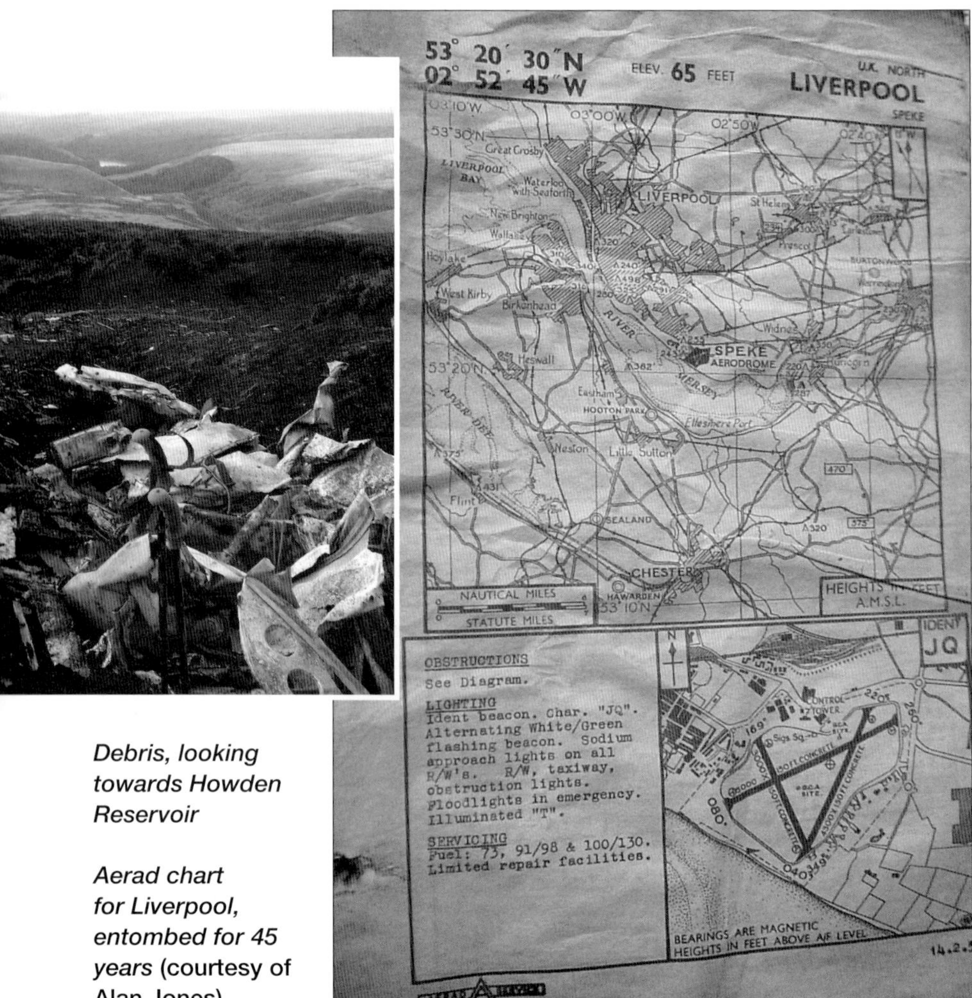

Debris, looking towards Howden Reservoir

Aerad chart for Liverpool, entombed for 45 years **(courtesy of Alan Jones)**

one hour and ten minutes at the Consul's 163 mph (140 knots) still-air cruise. The distance to the crash site they would have plotted at 160 miles, or one hour's lapse time; in planning terms, just when Captain Magnusson must have expected to be abeam Chester or thereabouts, and so a very reasonable time at which to think of edging down to re-establish visual ground contact before commencing an approach to land. That is, had the aircraft flown its planned track of approximately 320° rather than its more north-easterly actual track of 335°.

But what had taken it so far to the right of track? The extremely strong westerly wind blowing from the left? Or a compass – or compass-setting – error?

Turning away from what could only be speculation, the accident investigators found no sign of either structural or mechanical failure; indeed, they established that both engines had been under power when the aircraft struck. Their submission simply stated, therefore, that the pilot had strayed markedly off course, and after cruising above a cloud layer, and believing himself to be approaching Liverpool, had maintained cruising speed as he had descended blindly through cloud. Except that he had descended into the cloud-obscured high ground of a rock-strewn grouse moor, striking at nearly 1,600 feet above sea level when his altimeter – soullessly registering heights above the near sea-level datum to which it was set – would have persuaded him that he still had more than adequate clear air below.

In September 1996 researcher Mr Alan Jones unearthed a poignant survival from the 1951 crash in the form of the Aerad Company's radio facilities booklet which Captain Magnusson would next have referred to when making his approach to Liverpool. As Captain Magnusson had failed to get that far, the booklet remained as he had stowed it, enfolded between two newspapers; the three documents remaining perfectly legible for all their 45-year entombment in the peat.

Visiting the Site

The Consul crash site is reached from the Ladybower Reservoir access road running north from the A57 Snake Pass road. As far as the walker-driver is concerned the access road ends at the turning circle at King's Tree, for the gate beyond, even if unlocked on arrival, might well be secured. At least there is reasonable roadside parking just short of the turning circle.

To reach the Consul site, two miles off, takes about an hour (without refreshment stops), with some 225 metres (738 feet) to climb, mostly on well-defined paths. The walk begins by continuing to follow the road. Having crossed the Derwent by the re-sited stone bridge at Slippery Stones, and the Little Cut stream by footbridge, the most direct way to the site stays above the level of the Derwent to reach the mouth of Broadhead Clough. A substantial shooters' path then leads directly up the nearside side of the clough. Steep though it is, this ascent never becomes unremittingly tiresome, for it affords spectacular views throughout. Eventually, it climbs past some shooting butts, after which a track leads westwards to cross the clough and reach a parallel line of butts. In early 2010 a considerable amount of wreckage lay between butts eight and nine of this second row, the butts being numbered as they run uphill.

A worthwhile return route continues to climb to Howden Edge. It then turns eastwards along the rim path for two-thirds of a mile, providing the opportunity to take in the crash site of Oxford LX518 and two sites associated with a V1 and a parachute mine respectively (see below) before descending the Cut Gate path to Slippery Stones and King's Tree.

22

Airspeed Oxford Mk.1 LX518

Howden Moor, west of Margery Hill

SK 18025 96697	528 m, terminal site
SK 18043 96738	525 m, debris tumbled into a gully
Unit and Station	No. 21 (Pilots) Advanced Flying Unit, RAF Wheaton Aston, No. 21 Group, Flying Training Command
Date	18 October 1943
Crew	Pilot, killed Pilot Officer Dennis Patrick Kyne, Royal New Zealand Air Force

As they entered the advanced phase of their flying training, pupil pilots earmarked for a multi-engined role would be converted to one of the twin-engined trainers: the Oxford or the Anson. No. 21 (Pilots) Advanced Training Unit, then located at RAF Wheaton Aston, just south-west of Stafford, was equipped with Oxfords and for the initial night solo at this stage of the conversion it was the practice to carry out a short cross-country flight. The triangular course was flown first as a dual familiarisation detail, then as a solo training exercise.

Although integral to pilot training, night flying does make extra demands on the tyro pilot. This would have applied especially to one not native to Britain, and to one who, additionally, had learnt to fly under the Empire Air Training Scheme in the clear skies of uncluttered Canada. Accordingly, as they flew the dual exercise the instructor would refresh the pupil on the methods of gaining navigational assistance at night in the event of his becoming uncertain of his position when flying alone.

The instructor would demonstrate, for example, how a pilot could locate his position in relation to the coded 'pundit' beacons the active airfields would be showing; or how he could call into play occulting searchlights which, on request, would swing their beams to indicate the direction of an active airfield. Alternatively, if such lights were obscured for some reason, and the pilot was still unsure of his whereabouts, there was the 'Darky' network of listening stations which, by night, just as well as by day, could pass an aircraft from one to the other and so bring it to the nearest active airfield. Clear visibility, of course, was the most important aid of all, but to have proved the route just a short while before sending pupils solo eased the minds of authorising officers, instructors and pupils alike.

On the night of 18 October 1943, when Pilot Officer Dennis Kyne was sent off by himself, he had already flown the dual sortie. The route called for an 18 mile, south-westerly leg to the pundit light-beacon indicating Condover airfield; then a northerly 10 mile leg to the Shawbury pundit, probably then, as in later years, slowly coding 'SY' in morse (but only in peacetime, with no intruders on the prowl, actually located on the airfield itself!). The final leg would take Pilot Officer Kyne the 17 miles back to Wheaton Aston. It was an undemanding exercise which, according to plan, would see him back in the circuit in some 25 to 30 minutes. Only for the New Zealander things went sadly wrong. For Pilot Officer Kyne's aircraft did not return to base, and when it was eventually located it was found to have crashed on desolate moorland 52 miles to the north-east of Wheaton Aston.

The investigation would quickly discover that the Shawbury pundit, a beacon critical to Pilot Officer Kyne's navigation in marking the northern limit of his pattern, had become unserviceable only a short while after he had taken off. Which meant that on his northerly leg, searching ahead over otherwise blacked-out Britain, he would have seen only darkness instead of the reassuring pulse of light which had marked the turning point on the dual sortie. At his stage of training it must have seemed only sensible to fly on just a little more; after all, headwinds do spring up and alter estimated arrival times.

And things had certainly altered at Wheaton Aston, confounding the best efforts of the authorising staff to ensure that their pupils were not sent off in unduly adverse conditions; for just as the second wave of mainly solo pupils had become airborne, so the weather had deteriorated markedly. Most aircraft had returned to base as the weather closed in and one other, although initially causing some alarm, was found to

The terminal impact site, looking towards the Cut Gate path and Margery Hill

Debris which could not be burnt or salvaged was concealed in this gully; peat movement has brought it to the surface

have diverted, leaving just LX518 unaccounted for. After the prescribed lapse of time with nothing heard, 'overdue' action was taken as a matter of course, all agencies being contacted for information; but without result. Not until four days later was Pilot Officer Kyne's body discovered amid the wreckage of his Oxford, high on the Howden Moors.

Farmer Kenneth Wilson, of Greenwood Farm, near Hathersage, was able to throw light on the manner of its finding. 'We were beating for a grouse shoot on Howden Moor,' he said, 'when we came upon the wreckage, strung out over a half a mile or so. Then we saw the pilot, still strapped in his seat. I remember one of the farmers asking, as if in hope, "Is it a dummy?" But it clearly wasn't. After that, we stayed there, mounting a sort of guard while people went to raise the alarm, and brought back some servicemen.' He smiled suddenly. 'I remember they also brought us Spam sandwiches. And that I quite liked them!'

What caused Pilot Officer Kyne to stray so far north had to be a matter of speculation, but 52 miles was, after all, less than 20 minutes' flying time for an Oxford. Examination of the wreckage showed that Pilot Officer Kyne had not run out of fuel; accordingly the investigation turned to the procedures he might have adopted to get assistance.

It seemed clear that he had not made use of any radio facilities; although the 'Darky' posts in that area had been busy throughout the period, nothing had been heard of him. Further, his automatic alerting device, the 'Identification Friend or Foe' equipment (IFF), had not been switched to the correct channel; although that alone would hardly have saved him, even had a shepherd aircraft been able to locate him in time. As it was, it seemed that he had become totally lost, steering blindly north and north-east in a vain endeavour to fix himself. Except that, while unknowingly moving over higher and higher ground, he had maintained the height on his altimeter laid down for the low-lying Wheaton Aston area navigational route, until eventually his cruising aircraft had impacted on Howden Moor at an altitude of 1,700 feet above sea level, without catching fire but instantly killing Pilot Officer Kyne.

Crashed 18 October 1943

In 2010 a burn scar beside the footpath, caused by the fire set during the clearance operation and still liberally strewn with specks of molten metal, marked the terminal impact point. Additionally a fair amount of wreckage remained where the salvage team had tumbled it into a gully just a few yards to the north; like the burning, a precaution against its distracting future air searches.

Visiting the Site

Access to the Howden Moors is conveniently made from the roadside parking at the King's Tree turning circle at the northern end of the Ladybower Reservoir access road.

To reach the crash site of Oxford LX518 takes just over an hour with some two and a quarter miles to cover and 240 metres (800 feet) to climb, practically all on well-defined paths. The walk begins on the track running north from the turning circle. Having crossed the Derwent by the stone bridge at Slippery Stones, and the Little Cut stream by footbridge, the route keeps above the level of the Derwent until it reaches the mouth of Broadhead Clough. A shooters' path then leads directly up the side of the clough, at times being steep enough to call for halts to admire the scenery: fine hillsides and craggy skylines! Eventually, though, it climbs past some shooting butts, then flattens to cross a shoulder before finally ascending to the rim path of Howden Edge.

Just before reaching the shoulder, however, it is likely that a detour will be made to the west, to cross the now-narrow clough and visit the crash site of Consul TF-RPM (see above), nestling amid another line of shooting butts 300 yards away.

Having reached Howden Edge and turned right (leaving the 541 metre Outer Edge trig column to the left, though it is well worth a visit!), the terminal site of the Oxford is just yards away and actually on the Edge Path. The path, however, splits into so many strands hereabouts that it is possible to miss the patch of scarred earth altogether, although once seen the peppering of debris is very evident. On the other hand the terrain is such that the gully where the salvage team deposited the debris they could not reclaim, bury or burn, is not at all obvious, despite being only a few feet north of the path. The initial impact point, remember, was a full half-mile distant!

A quarter of a mile northish of both path-site and gully there is the waterlogged crater made by an air-launched V1 Flying Bomb (see below). To reach it requires making ground over trackless, but quite splendid, moor. Beware, though, for the crater's muddy floor can swallow an extended hiking pole without it bottoming! Another crater, made by a parachute mine, lies parallel to the path and some 700 yards further on. The easiest way to reach this, though, is to return to and follow the Edge Path until abeam the crater site then strike northish for 200 yards.

As for the descent to Howden Reservoir, half a mile on from the Oxford site the Howden Edge path, continuing south-eastwards, crosses that of Cut Gate. A 90-degree, hard right turn here – south-westerly – furnishes a fine, open-hillside descent to Slippery Stones and King's Tree.

25

FZG76 V1 (*Vergeltungswaffe* 1) *Fieseler* Fi103 Flying Bomb

Cut Gate, Howden Moor, north-west of Margery Hill

SK 18224 97083	520 m
Luftwaffe	Launched by Heinkel He.111s of KG53, Venlo, Holland
Date	24 December 1944

The German *Fern Ziel Geraet* (Long-Range Target Apparatus), their *Vergeltungswaffe 1* (Reprisal Weapon Number One), was a pilotless flying bomb with a range of some 150 miles, normally ramp-launched. It was powered by a singularly sounding pulse-jet engine which, at a preset range, cut out, the cessation of the sound indicating to the initiated below that an explosion was imminent.

The main V1 campaign started on 12 June 1944 and was sustained at high intensity, mainly against London and Antwerp, for 80 days. As early as 9 July 1944, however, as the Allied armies overran the launching sites in France, the Germans began to employ V1s launched from Heinkel He.111 bombers of *Kampfgeschwader* (bomber group) No. 53, initially operating from Venlo, in Holland. This meant that the V1s could strike targets in the Midlands; the Heinkel hosts, with the V1s carried under a wing-root, approaching the East Coast below the radar screen then briefly pulling up to 1,500 feet for the launch. This evasion measure failed, however, and fighter interceptions were so successful that the Germans were unable to maintain the offensive.

Just the same, a raid dispatched on Christmas Eve 1944 gave chilling notice of what the future might hold, with thirty-one V1s – one falling as far inland as Kelsall, near Chester – causing many casualties and some damage to property. During this wave of 'cruise-missile' attacks, the V1 featured here came down on remote moorland to the north-west of Margery Hill summit, on Howden Moor.

In September 2009 Mr David Appleyard, of Sheffield, drew attention to research by Mr Brian Thompson, of Stalybridge. The latter, by matching debris found in this deep, flooded and exceedingly muddy-floored crater with known Flying Bomb components, was able to show that it had, indeed, been caused by a V1. Groundbreaking new evidence, it seemed! Yet it was then apparent that in 1982 Mr Ron Collier (*Dark Peak, 2*, p.152) had actually photographed this particular crater but applied the co-ordinates

Crashed 24 December 1944

The Flying Bomb crater and (inset) the German parachute-mine crater

for a shallow crater lying a half-mile to the south-east.

In a similar matching of components Mr Thompson has established that this second, shallow and seasonally waterless – but still uncomfortably muddy – crater is, in fact, that of a German *Luftmine* B parachute mine, a 1,000 kilogram device triggered by a clockwork fuse.

Supplementing all this, in early 2010 research by Mr Geoff Eyre, tenant of the National Trust's Howden Moor, confirmed that neither crater existed when the pre-Howden Dam estate maps (c.1916) were drawn. Arguably of far more interest to most moorland walkers than metal bits and pieces, Mr Eyre is intimately involved with the Howden Moor heather regeneration project, the largest and most ambitious of such projects in the world!

Visiting the Site

Visiting – and descending from – this site would hardly be done in isolation, rather it would be taken in during a visit to the Consul and Oxford sites nearby (see above). The crater lies some 400 yards north-eastish of the Oxford site, across trackless but typically fine moorland. A visit in October 2009 detected a corroded and forcibly crumpled component, but nothing else. That being said, the crater, as a moorland feature, must be treated with caution, particularly by the lone walker. During the October 2009 visit, an extended hiking pole failed to find bottom through the glutinous mud and instantly dissuaded one such walker, now chillily divested of trousers and boots, from wading to the central hummock to extend his metal detector search!

Some 700 yards further south-east, parallel to the Howden Edge path, is the crater made by a German parachute mine (at SK 18586 96522 527m). The easiest way to reach this, though, is to return to the path, striking across the trackless moor for 200 yards when abeam.

Scraps of debris, displayed on the author's hat, 2010

24

Short Stirling Mk.3 LJ628

Upper Commons, east of Margery Hill, Howden Moor

SK 20291 95633	469 m, impact area
SK 20142 95520	480 m, terminal area, debris in gully
SK 21326 95900	361 m, wing fragments in Ewden Beck
Unit and Station	No. 1654 Heavy Conversion Unit, RAF Wigsley, No. 5 Group, Bomber Command
Date	21 July 1944
Crew	Ten aboard, two injured Squadron Leader Hadland, pilot, flight commander, observing Flying Officer O'Leary, screening pilot Flying Officer L. Gardiner, pilot under training Sergeant McDonald, navigator Sergeant James Coulson, bomb aimer, injured Sergeant John Gittings, screen flight engineer Sergeant Ludlow, flight engineer under training Sergeant Tex Burroughs, wireless operator Sergeant Lennox van Nierkirk, Royal Rhodesian Air Force, rear gunner, injured Sergeant Austin, mid-upper gunner, Royal Rhodesian Air Force

Stirling losses became so heavy compared with those of the higher-flying Halifaxes and Lancasters that in September 1944 it was removed from bombing operations; just the same it continued to serve on such duties as minelaying, glider-towing, electronic countermeasures, clandestine operations, and transportation. In addition, it was used for four-engined training prior to conversion to Lancasters or Halifaxes, with crews posted onto it being trained by No. 1654 Heavy Conversion Unit (HCU) located at RAF Wigsley, west of Lincoln.

One such crew, earmarked for Lancasters, which got airborne for its initial Stirling training sortie on 21 July 1944, was that gathered about him by HCU-trainee Flying Officer L. Gardiner. The normal crew complement of the Stirling was seven men, but even for a training flight this one was top heavy, with ten occupants, for in addition to the normal instructing – or 'screening' – staff, the unit's flight commander was aboard to carry out an overall monitoring.

The main gully, in early 2010, cleared of major debris

The sortie, being the initial introduction to the Stirling, was planned to be wide ranging. It started with a climb to height, followed by a fly-around during which various engines were stopped in order to familiarise the pilot and flight-engineer trainees with the technical problems and the altered handling qualities when either an outer or inner engine was lost. With all engines re-started, the aircraft was then descended preparatory to practising landings and take-offs back at Wigsley. However, in the descent, after 40 minutes in the air, having flown for the most part over complete cloud cover, and with everyone patently concentrating rather too much on what was going on inside the aircraft, a cry of alarm from the front turret changed the whole picture.

Looking up at the warning, the operating pilot saw, rushing towards him through the enveloping mist, what the bomb aimer in the front turret had seen just instants before, rough moorland where only sky had been expected. Moorland which, despite an instant application of full power and even brutish back pressure on the stick (see Glossary) so rapidly outclimbed the bomber that the aircraft struck heavily, and proceeded to disintegrate over a 200 yards' slide. Providentially, there was no fire.

In the immediate aftermath of the crash two separate pairs of crew members set off across the moors – one pair reaching Ewden Lodge Farm – and eventually brought assistance. Subsequently, although both the bomb aimer and the rear gunner received injuries which required surgery, all returned to flying duties. The bomb aimer, regaining fitness rather later than the gunner, joined another crew at the Lancaster Finishing School but later completed a tour of operations. The other trainees, having remained crewed together, also completed their 30 operations.

It became apparent to the court of inquiry that in the course of the sortie the aircraft had been allowed to drift far further west than anyone on board had suspected. Moreover, only the sketchiest attempt had been made to fix the aircraft's position before commencing the descent.

Crashed 21 July 1944

A mainwheel prepared for recovery, 2005 **(courtesy of The Stirling Project)**

Salvaged components, bound for preservation in 2005 **(courtesy of The Stirling Project)**

Similarly it was evident that during the engine-out-handling phase of the flight, and during the initial descent, the undue preoccupation of the pilots and flight engineers had led to a gradual and undetected frittering away of height. This was not a situation that would have been apparent to very many of the crew members, for each would have been busy with his own specialisation. Not only that, but had anyone noticed the altimeter, it would have shown an apparently healthy 1,500 feet or so, having been set to zero on the runway of RAF Wigsley where the elevation was just 70 feet above sea level. Certainly nobody aboard conceived that they had drifted over, and were now letting down towards, the Upper Commons area of Broomhead Moor, a moorland elevated to 1,560 feet above sea level!

There followed condemnation, from the officer commanding the unit upwards through every higher echelon, of all those members of the crew responsible for having omitted to acquire an adequate fix before beginning the descent. But although the Air Officer Commanding would rule, and his superior, the Air Officer Commanding-in-Chief, would concur, that the accident was 'caused by the disobedience [by the pilot] of orders for breaking cloud', the incident was clearly regarded as one of the hazards attendant upon intensive training; for it became a case of flying logbook endorsements in red, formal reproofs for both pupils and instructors – slapped wrists all around – then back to the task in hand.

In the 1980s former Sergeant John Gittings, the one-time screen flight engineer, met Peak Park Ranger Peter Jackson and spoke of that now far off moment of truth. 'I was in the astrodome,' he recalled, 'expecting the clouds to part and show us Wigsley's runways, when suddenly I felt this massive bump, and saw lumps of earth flying past.'

Mr Alwyn Haigh, of West Nab Farm, Bradfield, a schoolboy in 1944, remembered the crash well. 'It was a really foggy day,' he recalled, 'and for some time we'd heard this aircraft overhead. Eventually it came very low indeed, although it never broke cloud. Then we heard it had crashed.' Mr Haigh's first visit to the site had come just days later. 'Not long after we passed the Broomhead Moor Shooting Lodge we could see what turned out to be one of the wheels.' He paused. 'One of our souvenirs was a gyroscope, but although there were no sentries on the plane, the roads were guarded, and so eventually we ditched it.' Having reflected, he went on, 'What stuff the RAF wanted they took on crawler tractors over White Carr Ridge and the track to Cottage Farm, on the Mortimer Road, but they left loads at the site.'

Mr Lewis Couldwell, who was brought up in Garlic House Farm, remembered that several bits of wreckage were spilled off during the wartime salvage operation, with some unaccountably wandering off even that route. 'When we were out beating from the shooting lodge and had to drop down into Ewden Beck on our way to the northern side of the moor, we'd shelter from the rain under a wing that was propped over some rocks; though how it had got down there, we never did discover, for most wreckage they salvaged, but didn't trans-ship from Mortimer Road, was dumped onto Ewden Height – among the artillery targets, some of which they'd shoot at from over near North America Farm, beyond Range Moor Top, to the north-west.'

Until 2004 a great deal of debris was still strewn along the line of approach at the main site, whole panels flapping forlornly, and most notable of all, the great struts and vast smooth-tyred wheels. With the crash site being so remote, and the Stirling such a large aircraft, this very substantial amount of wreckage seemed destined to remain for all time. Early in 2005, however, 'The Stirling Project' preservation group obtained Service assistance to airlift the major items from the site for use as templates in the creation of a static-display Stirling. Just the same, enough debris remained in early 2010 to tempt the interested walker onto this otherwise largely unfrequented, and for the most part, trackless moor.

Walkers joyously heather-bashing across to this site, however, should remember that the moor was used as a Second World War firing range, and that any ordnance discovered will be potentially lethal.

Crashed 21 July 1944

Visiting the Site

Although this crash site lies in largely trackless heather moor it opens itself to several approach routes, parking for three of them being opportunity-roadside along the Midhopestones-Strines, Mortimer Road.

The most northerly approach is by the shooters' track which leads off Mortimer Road at SK 23880 93370, near Cottage Farm, before petering out after a mile or so to leave the walker free to pick a way across White Carr Ridge to Upper Commons. This route allows fine views of rolling moorland and is probably as straightforward as any. Certainly, vehicles used it to recover wreckage in both 1944 and 2005.

A similar route leaves Mortimer Road at SK 24180 96180, opposite Broomhead Hall, to give fine, uncomplicated moorland walking for the one and a half miles to the shooting lodge. After that, however, there remains over a mile of trackless moor where the run of the drainage is not at all friendly. It is probably a fair idea, therefore, to edge north and take in the segment of wreckage on the bank of Ewden Beck, near Gallow Rocher. Certainly, from then on, following convenient water channels towards Long Pole Ridge can considerably ease the way.

A third route off Mortimer Road, at SK 24555 94550, follows the Dukes Road Path for nearly two miles, to Flint Hill. After that, though, it is a matter of heather and grough tramping for the remaining mile or so. Good for the soul, perhaps, and a representative sample of the rough moorland walking which the gentry traditionally enjoyed, and paid through the nose for! But hard going, just the same.

A rather more dramatic approach, if merely exchanging laborious heather and bracken for severe gradients, is that from the northern end of the Ladybower Reservoir road. This starts at King's Tree and climbs the Cut Gate path to Howden Edge before turning south-east to Margery Hill. So far the path may have been steep, but at times it has even been paved! Beyond Margery Hill, though, lies a pathless heather bash of very nearly a mile, although from this direction the groughs are friendlier, if only by degree, than when approaching from the east. However, before adding this sortie to one involving the Oxford and Consul (see above) it should be borne in mind that the tramp back to Margery Hill is also a heathery moorland near-mile. But upslope!

Debris in Ewden Beck, early 2010

25

Vickers Armstrong Wellington Mk.1C DV810

Broomhead Moor, south-west of Stocksbridge

SK 23505 95465	377 m
Unit and Station	No. 21 Operational Training Unit (OTU), RAF Edgehill, No. 91 Group, Bomber Command
Date	9 December 1942
Crew	Seven on board, four injured Flying Officer Stanley Baker, OTU staff pilot Sergeant Anthony St Clair Turner, Royal Australian Air Force, trainee pilot, injured Flight Sergeant Donald Norman Dawson, navigator, injured Flight Sergeant Walter Samuel Sinclair, bomb aimer Sergeant Morgan, staff wireless operator Sergeant Alan Gordon Allwright, trainee wireless operator, injured Sergeant R.D. Weeks, air gunner, injured

The Wellington bomber, though swiftly outmoded by the heavier machines which later entered service, not only stayed the course but served with distinction throughout the Second World War. Always dependable, it also built up a reputation for being able to contain damage that would have crippled any other type; a reputation that relied almost entirely upon the singular geodetic framework developed by Barnes Wallis (see Glossary).

But that there were limits to even this ruggedness was to be shown on the night of 9 December 1942, when, together with five other Wellington crews, Staff Pilot Flying Officer Stanley Baker and his trainee crew got airborne from RAF Edgehill (north-west of Banbury), at that time a satellite of RAF Moreton-in-Marsh where No. 21 Operational Training Unit (OTU) had its headquarters.

Like many of his fellow instructors Flying Officer Baker was 'resting' between operational tours, and probably flying more often than he had ever done on his squadron. For the

Crashed 9 December 1942

The pilot in 1944, Wing Commander Stanley 'Tubby' Baker, DSO and bar, DFC and bar

OTUs were always hard pressed to provide sufficient crews to fill the gaps left in the ranks of the operational units, whether by war or natural wastage. In fact, the urgency had been illustrated that very night when, No. 21 OTU having submitted that deteriorating weather conditions would probably preclude landings back at Edgehill, the controlling No. 91 Group had ruled that diversions were quite in order once the exercises had been completed.

Flying Officer Baker's main task would have been to hover over the shoulder of Sergeant Anthony Turner, the trainee pilot – most Wellingtons having just the one pilot's seat – but as aircraft captain he would have held a watching brief over every member of the crew, and in particular over the navigator, upon whom so much depended.

Throughout the flight the navigator would have been busy at his lamplit table, poring over that esoteric mystery so beloved of the breed, the Air Plot. His divinations would be supplemented whenever possible by visual fixes from pilot, bomb aimer or gunner, and when exercise rules allowed, by bearings and fixes from the wireless operator; there were OTU exercises, more closely mimicking operational conditions, when wireless silence would be imposed, but this was clearly not one of those occasions, because Sergeant Morgan, a staff instructor, would have been specifically exercising the trainee wireless operator in his craft.

It was Sergeant Morgan who had the first intimation that all was not as it should be, reporting that they were picking up the hazard signals from a balloon barrage. Such warnings could only mean that they had drifted westwards from their supposed position and were now overflying some sensitive complex.

Cornish researcher, Mr Tony Meeks, at the impact site in November 2009

In this extremity Flying Officer Baker, a pilot with nearly a thousand flying hours, would have liaised with the qualified but inexperienced navigator, comparing notes and seeking to determine just how far astray an unseasonable easterly wind might have taken them. Eventually, forced to realise that they were quite lost, he exercised his command authority, exchanging seats with the trainee pilot preparatory to taking remedial action.

It is not clear why, especially with two wireless operators on board, radio assistance was not forthcoming, but this seems to have been the case. Nor does the short-range HF 'Darky' organisation appear to have been called into play from the pilot's seat; for although essentially a get-you-home service it could also furnish a rough but perfectly adequate position. But radio-aids aside, there were other self-sufficiency lost-procedures that fitted the bill, and it was upon one of these that Flying Officer Baker decided.

Crashed 9 December 1942

In the Midland area, over which they were most likely to have drifted, England is of the order of 140 miles wide. A distance which, at the Wellington's cruising airspeed of 173 mph (150 knots), translates as some 50 minutes' flying time; or rather more, allowing for the healthy easterly wind they would evidently have to head into. Accordingly, Flying Officer Baker set an easterly heading, away from the central spine of high ground and towards the Fenlands, the Wolds and the North Sea, for once safely over a flat area he could then afford to nose down, break cloud, and re-establish their position. After that, depending upon the weather conditions below cloud, he could return to base, or call it a night and divert. A workmanlike plan; except that he decided upon just 20 minutes as being enough to see them over terrain reasonably suited to making a blind descent.

True to his plan, exactly 20 minutes after setting course, Flying Officer Baker reduced power and committed himself and his crew to a blind descent. Any undue concern was allayed, no doubt, by the reflection that at least the altimeter, set with respect to the relatively low-lying RAF Edgehill, might be of some help, the Eastern Fenlands being low-lying themselves.

It would have been a reasonably tense descent in any circumstances, even for a crew positively certain of their position, with all eyes turned outwards and downwards. But with 1,200 feet on the altimeter it was Flying Officer Baker himself who first saw what he took to be yet another layer of cloud beneath him. Aided by a brief inspection with the landing lamp he decided that it was indeed cloud, then switched off the light again. After which, according to a statement attributed to him, 'Almost immediately there was a bang and my port engine was on fire.'

He then closed the throttles fully, hastily warned the crew of an impending crash and held off until the aircraft finally struck again.

It was providential that the terrain, though high-standing at 1,200 feet above sea level, was relatively level, for the aircraft slithered to a halt on its belly. Just the same, four of the crew were injured, and the instant all aboard had scrambled clear the machine erupted in flames. Worried that the fuel tanks might explode, Flying Officer Baker shepherded his shaken charges yet further clear, all seven being well distanced when, just moments later, the anticipated explosion actually occurred.

In the next hour or so, as the crew began to settle and take stock, the predominance of heather probably persuaded them that they were not in the Fenlands. Only it was not until locals reached them, drawn by the conflagration, that they discovered themselves to have come down on a Peakland moor just south-west of Stocksbridge, and some 35 miles short of any suggestion of reasonably low-lying ground.

The subsequent court of inquiry attributed most blame to the navigator; a puzzling finding, since for some time before the final descent all navigational decisions had been taken by Flying Officer Baker, the captain. As for Flying Officer Baker, he certainly suffered no lasting ill effects, but went on to have a most illustrious war, finishing up as the Wing Commander leader of No. 635 (Pathfinder) Squadron, with a well-earned double Distinguished Service Order, and an equally well-earned double Distinguished Flying Cross.

Mr Alwyn Haigh, of West Nab Farm, remembered the crash with guilty relish. 'I was still at school, and when we got to the Wellington, probably the next day, we found that although the front of it had burnt out the rear part was pretty well intact. So, of course, we made a beeline for the tail turret. Only when we started playing with the guns they let fly over the moors.' He grinned. 'At which we legged it away. Not that there were any guards, not on the moor. And like the Stirling crash at Margery Hill, the wreckage stayed there until it was dragged off.'

Visiting the Site

Limited lay-by parking is available on Mortimer Road (Midhopestones-Strines) at SK 24555 94550. The most convenient way to the Wellington site is to follow the well-established and undulating Dukes Road for just short of half a mile, a 20-minute stroll amid undulating heather moorland with fine views into Agden Clough. At SK 23499 94873, (where the now-accompanying, ruinous drystone wall kinks, and at a boundary stone inscribed PHRW), a northerly path leaves Dukes Road bound for a seasonally dry tarn. The crash site lies some 600 pathless yards – 20 minutes – beyond the tarn.

Having circumvented the tarn the approach route descends, dictated always by where the heather is easiest to tramp through. Certainly, given good visibility, a useful point to steer for is where the Mortimer Road crosses the crest of the distant hillside at Thorpe's Brow, above Cottage Farm (SK 23900 97480). Finally, just a few hundred feet before the crash site, a drainage ditch has to be negotiated, sizeable at four or five feet though nothing in comparison to the upper-moorland groughs.

The ground in the immediate area of the crash site is broken by tiny gullies bordered with lush heather growth, so it might well be necessary to search around regardless of the accuracy of the fix supplied. Even then, it is quite possible to be within feet of the debris before it reveals itself. In early 2010 there was still a moderate amount of wreckage, all of which had been transferred to a single debris pool at the terminal site. Researchers in the 1980s, it is held, were able to establish the point of the initial touchdown, but whatever scar there was has vanished long since.

On the ascending 20-minute return to Duke's Road, two singular, close-set boundary stones near the path provide a convenient aiming point, while the lacelike, ruinous drystone wall forms a backstop against the skyline.

Once re-established on the main path, the crash site of Hunter G-BTYL (see below) is about half an hour's walk away to the right – the west; that of Stirling LJ628 (see above) somewhat further and requiring even more seriously rough walking. But in proceeding anywhere off-path on this moor it should be borne in mind that the whole area has long been used for military exercises so that corroded, potentially lethal ordnance abounds.

26

Hawker Hunter TMk.7 G-BTYL (formerly XL595)

Brusten Croft Ridge, Broomhead Moor

SK 20873 94337	476 m
Operator	Cubitt Aviation
Date	11 June 1993
Crew	Pilot, killed Mr Wallace Cubitt

The Hunter being so effective a fighter, and such a joy to fly, it was hardly surprising that the RAF retained many in both operational and training roles long after the type was retired from first-line service at home. It is equally unsurprising that several of those released from the Service subsequently passed into civilian hands, among these being XL595, which, purchased by Cubitt Aviation and re-registered as G-BTYL, fatally crashed on 11 June 1993.

Limitations imposed by the Ministry upon the operation of this aircraft by the owner, Mr Wallace Cubitt, included the stipulations that it was not to be flown above 10,000 feet or in anything but good weather. In view of the height restriction, it hardly seems likely that Mr Cubitt would have gone to the extra expense and trouble of keeping the effectively-redundant oxygen system topped up and ready for use. Again, not being permitted to fly the Hunter above 10,000 feet meant that much of its greyhound-like performance had to be gallingly leashed in. All of which meant that there were heavy, officially-imposed restraints upon Mr Cubitt as he prepared to depart from RAF Coltishall to fly his aircraft to Preston's Warton airfield, where the fortieth anniversary of the Hunter was being celebrated.

He had been especially invited, and as Mrs Cubitt remembered on detailing the tragedy, was very keen to make the flight. However, on repairing to the met office at 0945 hours he found that heavy build-ups of turbulent cloud precluded the filing of a clear-weather flight plan. Disappointed, Mr Cubitt kept checking periodically, until 1350 hours when he made a final visit. On being told that the weather conditions had still not improved, he nodded and clearly dismayed, muttered, 'I'll have to think about it.' Accordingly, having thought about it, and without doubt considering that the actual conditions were such that

Debris gathered beside the impact crater, April 2010

he stood a good chance of finding gaps enough to enable him to complete the flight in the clear, he made his very proper decision and got airborne at 1431 hours. Only not long afterwards his aircraft, demonstrably out of control, dived into Brusten Croft Moor at very high speed, burying itself in a deep, peaty crater.

Researcher Alan Jones was an early visitor at the site, arriving just a day after the event. 'The army had taped the place off,' he remembered. 'They had all-terrain vehicles and pumping equipment and made great efforts to reach the cockpit. Their pumps, however, were unable to keep pace with the water and finally they were forced to give up, leaving Mr. Cubitt interred with his machine.'

In part because of this inability to recover so much of the wreckage, and in the absence of any radio report regarding mechanical malfunction, determining the cause of the accident had to be somewhat speculative.

One line taken was that Mr Cubitt had entered cloud, been obliged to resort to flight by reference to instruments alone, and either from disorientation or from the effects of turbulence, had lost control; after which the Hunter, being such an eager machine, would have lost height with extreme rapidity. On the other hand, although Mr Cubitt had flown only eight hours in the Hunter, the investigation recorded that he had 5,600 hours overall, an experience level which speaks for his probable competence in instrument flying (except that his currency – that so-vital factor where instrument flying is concerned! – was not specified).

Crashed 11 June 1993

Another line pursued was that, in seeking to find clear weather, Mr Cubitt had climbed above the limiting 10,000 feet to an altitude where he could pick his way between the tops of build-ups. Doing so, however, would have been fraught, for by 25,000 feet, without supplementary oxygen or a pressurised cabin, and with their faculties being gradually diminished by lack of oxygen, the average person might expect to lapse into actual unconsciousness. Of the two, it has to be said that the latter cause, loss of control due to lack of oxygen – hypoxia – seems the more likely.

In early 2010 the crater gave every appearance of remaining a prominent feature of the moor for many years to come, the surrounding area being silvered with tiny fragments of surprisingly thick-gauge metal literally torn apart by the force of the impact. All the more of a shame, then, that sections of the perforated steel plate used by the recovery team had been left protruding from the steadily gathering water.

Visiting the Site

Limited parking is available on Mortimer Road (Midhopestones-Strines) at SK 24555 94550. There follows a two and a half mile (45 minute) walk along the Dukes Road track to SK 21135 93983 where a boundary stone, to the left, is inscribed RRW. The site is then some 480 yards, 327°M, from the track over pathless and rough, but relatively unchallenging, terrain. Unchallenging, that is, provided it is borne in mind that any ordnance discovered should be treated as potentially lethal!

The discerning walker will not, of course, fail to visit the unnamed tarn nearby, at SK 20919 94140, a fine picnic spot. For those specifically visiting crash sites, the route passes abeam of Wellington DV810 (see above), some 20 minutes after leaving the car. The crash site of Stirling LJ628 (see above) is also within reach, though this requires even more very rough walking.

A small selection of the lethal ordnance recovered from the moors by Mr John Ownsworth

27

Gloster Meteor FMk.4 RA487

Hagg Side, above Ladybower Reservoir

SK 16594 89066	346 m
Unit and Station	No. 66 Squadron, RAF Linton-on-Ouse, No. 12 Group, Fighter Command
Date	8 December 1950
Crew	Pilot, baled out safely Sergeant Joseph Harrington

After the Second World War the Meteor formed the mainstay of Britain's fighter force until replaced by the new generation of swept-wing jet fighters, notably the Sabre, which No. 66 Fighter Squadron, among others, were to receive in 1953. But in 1949, when the squadron was still Meteor-equipped, they were joined by Sergeant Joseph Harrington who had re-enlisted in the RAF after a year's absence following late-wartime service on Lancasters and Lincolns. Having quickly settled into this jet-fighter environment, Sergeant Harrington had already become a member of the squadron's Meteor formation team when, towards midnight on 7 December 1950, he was dispatched on a triangular cross-country exercise.

The first leg of his route, to Manchester, was uneventful. Shortly after turning onto an easterly heading for his second leg, however, he realised that his communications radio, the state of the art, four-stud, very high frequency (VHF) set, had totally failed. Accordingly he decided to adhere to his navigation plan, to turn north for base at his estimated time of arrival for the final turning point, and then commence his descent. When the time came for the turn, though, and being very conscious of the safety height along the leg, he found himself above a substantial, if occasionally broken, cloud layer.

Reassessing the situation in view of the cloud, with fuel running relatively low, and loath to let down before more accurately determining his position, Sergeant Harrington switched his 'Identification Friend or Foe' equipment to 'Emergency', knowing that this would show up as a distinct 'Help me!' signal on the fighter-control radar screens. He might also have considered flying the arcane emergency procedure known as 'radar triangles', in this case to the left, hoping that the radars would recognise such peregrinations as indicating 'My radio has totally failed, prithee send up a shepherd

The pilot, Sergeant Joe Harrington

aircraft to see me home'. But optimism has its bounds for anyone, and with his fuel state increasingly in mind, Sergeant Harrington decided to take advantage of a chance cloud break and investigate the possibility of making a precautionary landing.

Edging down through the gap, maintaining what visual contact he could, he actually succeeded in picking out a straight stretch of road through the darkness. In fact, he even made a tentative pass along it with a touchdown in mind, only to be compelled into a hasty zoom for height as a seemingly sky-tall chimney loomed off one wing.

Having ventured so much, however, Sergeant Harrington then did what so many others lost in the region had signally failed to do. Deciding that enough was enough, and before his fuel state further cut down his options, he made the balanced decision to leave the machine to its own devices.

Not that this was to go without a hitch, for having climbed back to an indicated safe altitude of 5,000 feet, and having suitably configured the aircraft, his initial attempt to evacuate was foiled by the forces involved. Indeed, only after laboriously pulling himself back into his seat, and more advantageously trimming the controls, was he able to spill himself over the side and thereafter – Fortune herself remaining with him! – under the tail. That accomplished, however, his parachute descent was successful and he landed, somewhat shocked and shaken but otherwise uninjured, in a ploughed field near Castleton. His abandoned aircraft, meanwhile, crashed on the isolated slopes of Hagg Side, high above the A57, Sheffield to Glossop road.

Station Officer Roe, of the Glossop Fire Service, told the *Glossop Chronicle* how his crew had been forced to leave their fire tender at Rowlee Farm and Land-Rover their way along tracks to Lockerbrook Farm, eventually getting to within 350 yards of the machine, but only locating the wreckage having tortuously scrambled through thick undergrowth into the trees of Hagg Side Wood. The fire had gone out by the time they reached the site, but it was clear to all that chance alone had prevented the whole plantation from going up in flames.

Those junior RAF officers enquiring into the loss of the aircraft opined that Sergeant

Harrington might have either more accurately planned, or flown, the second leg of the route. Also that he might have realised earlier that he had lost his radio. And no doubt they had something in mind, albeit unstated, regarding what he might have done then. More mature Higher Authority, however, decided that matters need go no further, and it was left to the station commander to discuss with Sergeant Harrington the errors the inquiry felt he might have avoided. But clearly the miscreant's cool-headed decision-making was to stand him in good stead, for he subsequently made a career as a captain with Swissair; a premier airline reputedly demanding very nearly the same Teutonic levels of efficiency envisaged by the RAF court of inquiry, only supplying dependable radios, and paying a very great deal more handsomely.

Ranger Peter Jackson first met former Sergeant Harrington at a Glossop book launch in the 1990s. 'I fly into Manchester in my Swissair jet,' Captain Harrington told him, 'and often get the chance to look down at the area. But this is the first time I've revisited since floating into it on my 'chute.'

Visiting the Site

The Meteor site can conveniently be visited from the A57 Snake Road at SK 16259 88610, the nearest restricted lay-by parking area to the Hagg Farm track. Beyond the Hagg Farm driveway, the track zig-zags up to the corner of wooded Hagg Side where, at SK 16388 89072, a stile gives access to a substantial footpath (not presently marked on the map) running generally easterly. Following this to SK 16588 89036 the crash site is just 30 yards off to the left, amid the trees, a little short of a drystone wall.

If the ascent is made from the Ladybower Reservoir road near Fairholmes, parking is no problem. The track starts just south of the Information Station and leads up past Lockerbrook Farm to run southerly along wooded Hagg Side.

Having left the site and walked back to the zig-zag, then heading north-west affords a bracing and untaxing walk on Rowlee Pasture towards the must-see Alport Castles two miles away. Of course, after a mile or so, a detour might be made to take in the crash site of Defiant N1766 (see below).

The debris pool, inside Hagg Side Wood, in early 2010

Crashed 8 December 1950

28

Boulton Paul Defiant N1766

Rowlee Pasture, west of Derwent Reservoir

SK 15286 90495	448 m
Unit and Station	No. 96 Squadron, RAF Cranage, No. 9 Group, Fighter Command
Date	13 April 1941
Crew	Two, baled out safely Flight Lieutenant Paul W. Rabone, New Zealand, pilot Flying Officer John Ritchie, air gunner

By August 1940 the Defiant's mediocre performance had led to its being withdrawn from daytime operations. When re-employed as a night-fighter, however, it did rather better, and it was during this second phase of its operational life, on the night of 12 April 1941, that Flight Lieutenant Paul Rabone, a New Zealander serving with the RAF, was detailed to carry out a navigational exercise. The round-trip route was planned to take him and his gunner from RAF Cranage (seven miles south-east of Northwich, Cheshire) to RAF Digby (north-east of Cranwell, Lincolnshire), and return.

The outbound leg to Digby was uneventful but on the return, when the cloud lowered, Flight Lieutenant Rabone, being mindful of the region's high ground, increased his altitude and called for a radio homing to Cranage. He received no intelligible response, however, and although he persisted in trying to make contact, the wireless set eventually died altogether. On top of which he was becoming increasingly concerned about the engine; indeed the subsequent inquiry, exonerating him from all blame in the loss of the aircraft, records that it actually cut out. But evidently before that could happen, and having discussed the situation with his air gunner, the decision was taken to bale out.

There was no other sensible option. They were lost, in the dark with a wartime blackout below, deprived of radio assistance, uneasy about the engine, above solid cloud, and with hungry high ground waiting should they seek to penetrate that cloud in a blind descent. Accordingly, as Flight Lieutenant Rabone subsequently stated in his report, he climbed the machine until his altimeter indicated 3,000 feet, then ordered Flying Officer Ritchie to abandon; after which both of them made successful parachute descents.

Crashed 13 April 1941

The debris pool in early 2010

Flight Lieutenant Rabone himself was no stranger to parachuting, having previously baled out five times from both Fairey Battles and Hawker Hurricanes. Flying Officer Ritchie, on the other hand, was a novice jumper, but he too landed safely. On seeing a light, however, and making his way to a farmhouse, he had some difficulty in convincing the shotgun-wielding farmer that his thick Scottish accent was not German.

It was to Flight Lieutenant Rabone's credit, then, that he did not fall into the only too common trap of 'pressing on regardless' and of descending blind when lost above high ground. Nor, despite this being his sixth abandoned aircraft, was there any doubting his mettle, for he had already proved himself a successful – and enterprising – pilot and by the time he went missing in July 1944, during a cross-Channel sweep on Mosquitoes, he would have had ten enemy aircraft attributed to him.

A reasonable amount of wreckage remained in 2010, although an excavation was mounted in 1980 to remove the engine, said to be on display at the South Yorkshire Air Museum in Doncaster.

Visiting the Site

From the A57 Snake Road, three 220 metre (730 feet) ascents offer themselves for visiting the Defiant site on Rowlee Pasture: via Hagg Farm, Rowlee Farm and Alport Castles Farm. In all cases parking is the problem, although roadside parking does exist.

Ascending from both Hagg Farm and Rowlee Farm leads to the well-defined north-westerly track which runs across the spacious Rowlee Pasture towards Alport Castles. From the track the area of the crash site is hidden by a rise in the terrain, but in the vicinity of SK 15228 90049, a discernible footpath leads across rough meadowlands to pass close to the debris pool after some 500 yards.

The alternative ascent via Alport Castles Farm is equally worthwhile as a descent, but in either case no walker would want to leave the Pasture without having slightly extended the upland walk and taken in Alport Castles and The Tower.

Better parking is found at the Fairholmes Information Centre on the road along the Ladybower Reservoir. In this instance the ascent to the Pastures is made via Lockerbrook Farm. The woods through which the path passes, however, tend to cut down the otherwise magnificent views.

Both the Hagg Farm and the Lockerbrook Farm approaches lend themselves to visiting the Meteor site (see above), just yards within Hagg Side Wood.

29

De Havilland Vampire FB Mk.5 WA400

Strines Moor

SK 21842 89801	346 m
Unit and Station	No. 102 Refresher Flying School, RAF Finningley, No. 25 Group, Flying Training Command
Date	25 July 1951
Crew	Pilot, injured Flying Officer Lawrence Leslie Beckford

The de Havilland Vampire entered RAF service in 1946 but until 1950 no dedicated trainer version existed, the Mk.5 Fighter-Bomber variant being used for the task. One of the shortcomings of this single-seater stop-gap, however, was that the pupil pilot was not afforded the opportunity to sit beneath a visor, with an instructor to serve as both mentor and lookout, while building up experience in flying solely by reference to instruments. Further, although the flight instruments in the Vampire were well up to the standards of the day, there were certain foibles associated with the Mk.4F compass that could quite easily catch out the inexperienced or the unwary. A stand-by compass existed – the E2-type – but this does not seem to have been fitted to all Vampire aircraft. Besides, the diminutive E2 always gave the appearance of having come from a Christmas cracker, and rarely having to be used 'for real' was not always deemed worthy of attention, for all that it would be fitted to Concorde!

Another equally fundamental shortcoming of the Mk.5 Vampire was the lack of an ejection seat, for a manual abandonment was very likely to result in the pilot striking the tailplane. Certainly, had he had an ejection seat, Flying Officer Lawrence Beckford, a qualified pilot who had rejoined the Service and was re-acclimatising to flying at the Refresher Flying Training School at RAF Finningley, might well have considered ejecting, rather than forced-landing on rough moorland when he lost himself and then ran short of fuel on 25 July 1951.

Flying in Vampire FB Mk.5, WA400, Flying Officer Beckford had been airborne for an hour and 30 minutes on a daytime solo sortie during which he had been authorised to carry out a series of high- and medium-level exercises including a high-speed, or 'compressibility', run close to the machine's limiting Mach number: of the order of 0.85M (see Glossary). The high-speed run, starting at 40,000 feet, would have been straightforward – always supposing that WA400's airframe had not been overly brutalised by student-handling in the past – but Flying Officer Beckford had also been authorised to carry out aerobatics. And it may be assumed that he had carried these out with full verve.

Then came the recovery to base, starting off with the call for a homing to the Finningley overhead and finishing with him being deposited below the cloud base – real or simulated – and on a heading that would have allowed him to join the circuit visually, and land. But prior to all that, after his aerobatics and even before calling for the homing, Flying Officer Beckford should have flown straight and level while re-erecting his gyroscopically-controlled artificial horizon, and while carefully re-setting and double-checking the synchronisation of his equally gyroscopically-stabilised compass. Only it is just as likely that he cheated a little, as most people did, and busied himself with re-setting his instruments during the wings-level preparatory descent; straight, certainly, and if not *quite* level – well, level-ish.

This likelihood is given weight by the fact that after Finningley had told him the direction to steer for their overhead, subsequent R/T exchanges showed that he was heading away from the airfield rather than towards it. Naturally, this would have caused mild consternation on the ground, but it was initially assumed that he had simply misunderstood the instructions. What seems likely, however, is that in his hurry to descend Flying Officer Beckford had indeed paid insufficient attention to synchronising his gyro-magnetic compass, which was consequently giving him a false heading indication. It was an error only too easy to make on the compass in question, when it was merely a case of twisting a knob towards either a dot or a cross – and towards the wrong symbol, as often as not, if half occupied with doing something else.

The author at the crash site of Vampire WA400 in September 2004; finally rediscovered thanks to information supplied by researchers John Ownsworth and Alan Jones

Chagrined as this ex-operational-squadron pilot must have been at losing himself like a novice, but with almost total cloud cover preventing him from getting a visual fix, Flying Officer Beckford would have realised that, being also low in fuel, he had just two options. To gain height, and using the risk-laden technique recommended for safety, abandon the aircraft. Or to descend through a gap, and put the aircraft down as best he could. Of the two he chose the latter.

In fact, he broke cloud near Strines Moor, in the vicinity of the celebrated Strines Inn, and was recognised as being in trouble by Mr Wilfred Livesey, an off-duty ambulance driver who lived in a cottage in the grounds of nearby Sugworth Hall and was out walking with his five-year-old daughter, Jill.

As Jill, subsequently to become Mrs Arnall, recalled, 'We were half-way up the drive to the road and the aeroplane flew over our heads, going to the right of Bents House. What I always remember was the noise – the screaming sound. But daddy knew it was going to crash, so he ran back and phoned the fire service. Then he drove round by the cattle grid – it's tarmacked over now – and just where the road kinks, as you approach the Strines Inn, saw the aeroplane on the moor. So he ran up to it, and was just in time to help the pilot away – he'd grazed his knee, that was all – when it caught fire. Then he brought the pilot back to the Hall.'

Whatever the causative problem, the compass, together with the rest of the aircraft, was totally consumed by the flames. But Flying Officer Beckford had clearly drawn upon each of his six hundred or so hours of flying experience by choosing an only-gently ascending moorland slope to alight on. Further, having approached at the prescribed 140 knots, albeit choosing to land wheels-up, he had finally had the luck he deserved, for in the course of the aircraft's slide to a halt it had hit none of the moor's ubiquitous boulders. As for his choice of landing wheels-up, at that time Vampire *Pilot's Notes* directed a forced-landing pilot to select the undercarriage down 'unless the terrain is unsuitable'; not for some years would an ammendment alter this to 'regardless of the terrain'.

The aircraft having burnt out, both the court of inquiry and Higher Authority seem to have flailed around rather. While noting that Flying Officer Beckford had only 52 hours of jet time, they accepted that the aircraft compass had become unserviceable, but loftily observed that, as the cloud cover had not been total, Flying Officer Beckford should have deduced from the sun the direction he was flying in. (True, nothing was said about moss, and the northern side of trees.)

Then Air Traffic were castigated for not having ordered a 90-degree turn, duly noted the change in bearing, and calculated from that what heading the pilot was actually on. Finally, it was the turn of the Equipment Branch, who, with more credibility, were enjoined to see that E2 compasses were henceforth fitted as a standby to all the Command's aircraft whose primary compass was the G4-type.

For many years the location of this crash site defied detection, particularly as the popular printed sources were so far out, but in September 2004 aviation artist and

Crashed 25 July 1951

The debris pool in early 2010

researcher Mr Alan Jones supplied a reference from researcher Mr John Ownsworth which led to its re-discovery. In fact, it lies some 30 miles from Finningley, on a hummocky lift of heather-covered and boulder-strewn grouse moor belonging to the Fitzwilliam (Wentworth) Estates but – since the September 2004 'Right to Roam' legislation – normally open to the public.

Visiting the Site

The crash site is located just off Mortimer Road (A57 at Moscar north to Midhopestones) and about half a mile south of the Strines Inn. Lay-by parking is available at SK 22137 89929, on the south-east side of the road, but as this is close to a blind hill summit, care is called for. Access to the moor – a managed grouse moor – is diagonally across the road, via a gate. From this gate, the site lies on a heading of 245°M at a distance of 380 yards. The going is rough across heather, and essentially featureless. But 143 feet (by GPS!) before the site is a lone, large boulder. In early 2010 all that remained were some sheets of molten metal. The heather growth, therefore, could well make it necessary to search about from the reference given.

30

De Havilland Vampire T.11 XE854

Occupation Road, Rawmarsh, Parkgate, Rotherham

SK 43356 95747	37 m
Unit and Station	Fleet Air Arm, No. 1 Advanced Flying Training School, RAF Linton-on-Ouse, No. 23 Group, Flying Training Command
Date	9 March 1959
Crew	Pupil pilot, killed Midshipman Ian Ferguson Wilson, Fleet Air Arm

At 1047 hours on 9 March 1959, Midshipman Ian Ferguson Wilson, a pupil pilot being trained for the Fleet Air Arm at RAF Linton-on-Ouse, York, took off in a Vampire T.11 ('tee-eleven') to practise circuit flying at Linton's relief airfield, Full Sutton. Having carried out a number of take-offs and landings, as the duty entailed, Midshipman Wilson cleared the Full Sutton area and declared his intention of changing to Linton's radio frequency.

He failed to check in with Linton, however, and when a standard 'Airborne forty minutes' advisory was transmitted to him by the tower, there was no response. In fact, at 1120 hours he had crashed his Vampire some 30 miles from Linton, after low flying near his home on the outskirts of Rotherham. Midshipman Wilson had been killed as the machine exploded, but providentially nobody on the ground had been hurt; and although there had been some disruption at a nearby colliery as the aircraft severed power cables, there had been little damage to property.

The incident was covered in predictably journalistic fashion by the local press – 'Hero pilot', and the like. The coroner's investigation on 28 March 1959, however, was far more pertinent. Quizzing Midshipman Wilson's uncle, Mr James Ferguson, of Rotherham, the coroner observed, 'When he was killed he had been flying...more or less over his home.' He then asked, 'Do you know if he had done this on any previous occasions?' Mr Ferguson replied ingenuously, and damningly, 'Oh yes! Several times recently, but in a piston-engined plane. This was the first time in a jet.'

The coroner also heard evidence from Full Sutton's Duty Air Traffic Controller, and from Midshipman Wilson's squadron commander. On being told that the pupil pilot had been briefed to 'do five circuits, then five rollers, then return,' (see Glossary) the

(top) The overgrown memorial site, as late as November 2008 and (above) the new memorial unveiled in 2010

coroner asked, 'Would these exercises take him over Rotherham?' 'No,' the air traffic control officer replied unequivocally, 'that would take him about 20 miles outside his flying area. He would have known that.' The squadron commander was equally blunt: 'It was an act of indiscipline.'

Describing the crash itself, Mr Raymond Wilkinson, of Rotherham, told the coroner, 'As [the plane] passed over the church it seemed to be losing height ... just above roof level. It made a quick turn away from the houses, and disappeared from sight in cloud. [On diving out of cloud again] there were green flashes as it broke through the power cables, then there was a loud explosion.'

The RAF board of inquiry would have had no need to deliberate over-long; Midshipman Wilson had been illegally low-flying and had been seen leaving his assigned area to do so. After that, witness evidence, professionally interpreted, told the story. He had begun a 'hairpin' turn-back (see Glossary) for a second run over his home, but on unexpectedly encountering cloud while in this steeply banked climbing turn at very low level, had become disorientated and lost control, his aircraft spiralling into the ground. Case closed.

In 2006 Mr Alan Smith, of Rawmarsh, described the scene on the evening of the crash. 'The Vampire had dived into the hillside and disintegrated, so that bits were spread back from the crater, over the rail line, and up the far side. I picked up an instrument from the cockpit; there seemed nothing any bigger left. It was gruesome, too ...' He reflected for a while. 'A memorial was built later, rather like a well, but it's years since I saw it, and it was overgrown even then.'

And a memorial there was. Once located. Raised by the family, it took the form of an octagonal enclosure two or three yards across, enclosed by a brick wall some three feet high. A decorative wrought-iron gate opened onto a centrally-located plinth beyond which, set beneath the raised back wall, was a wooden bench. There was also a double stone erected by other members of the family. It transpired that the plot had been donated by the owners of the slope, Mr and Mrs T. Fieldhouse of Parkgate; the memorial, centred over the impact crater, being opened on 18 September 1959, when Midshipman Wilson's ashes were deposited beneath the plinth.

Visiting the Site

In November 2008 the memorial was still overgrown, but by March 2010 local initiatives had operated to have a new memorial located in the town's Second World War Memorial Gardens (at SK 43578 95884). Just the same, the local newspaper, *The Star*, conformed to type, recording that the 'pilot ...(had) crashed his jet trying to avoid casualties', that he 'was killed trying to land the plane safely' after it 'developed technical problems'.

Non-Debris Sites

1
FZG76 V1 (*Vergeltungswaffe* 1) *Fieseler* Fi103 Flying Bomb

Westwood Farm, Matley

SJ 97550 95290	166 m
Luftwaffe	Launched by Heinkel He111s of KG53
Date	24 December 1944

One of the V1 pulse-jet Flying Bombs air-launched against the Midlands on Christmas Eve 1944, landed beside Westwood Farm at Matley, near Hyde, Tameside. Striking a clump of trees, it exploded, demolishing the farm and killing two of the occupants.

2
De Havilland DH9 F2751

Lower Mudd Farm, Mudd, Mottram

SJ 99590 94696	223 m
Unit and Station	No. 38 Training Depot Station, RAF Tadcaster
Date	18 January 1919
Crew	Pilot, uninjured Second Lieutenant William Henry Carr Robson, RAF

On 18 January 1919 Second Lieutenant William Henry Carr Robson was flying de Havilland DH9 F2751 when its engine began to cause such concern that he put it down at Mr Joseph James' Lower Mudd Farm, Mottram.

An RAF guard was provided, comprising Corporal Walter Jennings and Aircraftman Second Class Percy Ruffle, who were granted billeting facilities at Parsonage Farm, Mottram. At about midnight Corporal Jennings, concerned over the protracted absence of his relief, left his post at the site, and on arriving at the farm, found that Aircraftman Ruffle had dozed off after his supper break. Very shortly afterwards Mr James' son, Fred, bearing fresh burn injuries, arrived to report that the aircraft was ablaze. The machine was found to be beyond saving.

The newspaper reports of the subsequent proceedings were unearthed in 2006 by the independent diligence of Mr Arnold Willerton, of Hyde, and Mr Bill Johnson, of Mottram. In the face of seemingly conclusive evidence, Mr Fred James was charged with stealing 30 gallons of aviation spirit, valued at £5, and causing malicious damage to an aircraft valued at 'over £2,000' while attempting to siphon off fuel by the light of a storm lantern. After a court hearing, however, the charges were dismissed.

The map reference for the occurrence given here (just off the Littlemoor Road) was based upon courtroom testimony which located the aircraft in the 'next-but-one field to Lower Mudd, but out of sight of the farm'. In 2006 a rather cursory metal detector scan was carried out during a site visit made in company with artist and researcher Mr Alan Jones, but nothing was found; nor was anecdotal evidence forthcoming at the farm or elsewhere.

Miles Master Mk.3 W8474

Warhill, Mottram in Longdendale

SJ 99773 95325	221 m
Unit and Station	No. 16 (Polish) Service Flying Training School, RAF Newton, No. 21 Group, Flying Training Command
Date	19 July 1942
Crew	Pilot, killed Pupil Pilot Leading Aircraftman Józef Gawkowski, Polish Air Force under British Command

The Miles Master was an advanced trainer whose Spitfire-like characteristics were a joy to pupils graduating from the lower-performance Tiger Moths. But it also tested their powers of self-discipline. On 19 July 1942 pupil-pilot Leading Aircraftman Józef Gawkowski of RAF Newton, near Nottingham, was briefed for a local area training flight. However, as the RAF investigation summary records, he 'disobeyed orders and flew North, circled a village three times, then banked steeply, stalled his aircraft, which then went into a spin, crashed, turned over, and caught fire.' As the Air Officer Commanding commented, rather wearily, one imagines, 'Yet another case of the pilot disobeying his instructions'.

In March 2006 a memorial (see page 135) was placed at the entrance to the cemetery to the memory of – as the inscription's drafter saw it – 'this dashing young pilot'. The memorial itself was the fruition of 30 years of heartfelt campaigning by Mr Alan Jones, the locally-born air-crash researcher and a talented artist whose painting of the event is reproduced on the plaque in medium-relief. The gully is easily accessed from a footpath but all that might be found of this needless tragedy would be cinder-like soil.

4

Bristol Blenheim Mk.1F K7172

Woolley Bridge, Glossop

SK 00395 95661	135 m, touchdown area
SK 00480 95555	128 m, termination point
Unit and Station	No. 29 Squadron, RAF Digby, No. 12 Group, Fighter Command
Date	3 December 1940
Crew	Pilot, uninjured Pilot Officer Donald Anderson

While flying Blenheim K7172, Pilot Officer Donald 'Don' Anderson suffered a radio failure as he neared the end of a communications trials sortie. With the light fading and his fuel state giving concern he carried out a blind descent through cloud, emerging in a wide valley at Woolley Bridge, Glossop. Unwilling to press his luck further, he set himself up for a precautionary landing. As he approached, however, he realised that the ground was sloping away so steeply that, despite full flap, it was some time before he could put his wheels on the ground and begin killing his speed. As he later conceded, his hasty braking then caused the Blenheim to flip over onto its back. Pilot Officer Anderson was able to release his straps, tumble onto the roof without breaking his neck, and extricate himself. The machine, though, had to be scrapped.

The court of enquiry felt that with just under 300 flying hours Pilot Officer Anderson should have handled the incident more competently. In consequence, an endorsement of 'lack of judgement' was duly entered into his flying log book.

Prospective visitors will find extremely limited parking at SK 00810 95780, on the A57. The Tameside Trail footpath leaves the road to pass through private property, after which a footpath branching off for Mottram leads to the site. There is nothing to be seen, however.

5

Fairey Swordfish, unidentified

Cross Cliffe, Glossop

SK 04021 94033	83 m
Service	Royal Naval Air Service
Date	June-July 1945
Crew	Pilot, uninjured

The Swordfish, as Mr Derek Slack of Glossop remembered, was 'in grey, Naval camouflage and seems to have got lost and run out of fuel. It touched down parallel to Hurst Brook, where Slant Close now leads off Shirebrook Drive. It then sat there for some days before the wings were taken off and it was loaded onto an RAF low-loader trailer.'

Mr Jim Buie, Secretary of the Greater Manchester Fleet Air Arm Association, was able to proffer that between June and July 1945 scores of Swordfish were being flown into Barton (Manchester) for scrapping. It rather looks, therefore, as if this particular veteran had decided to depart with rather more dignity than those who condemned it to the scrapyard had intended.

North American P-51D Mustang 44-64084

Plainsteads Farm, Monk's Road, near Glossop

SK 02620 91369	337 m
Unit and Station	United States Eighth Army Air Force, 2nd Air Division, 4th Fighter Group, 336th Fighter Squadron, AAF356 (RAF Debden)
Date	29 May 1945
Crew	Pilot, killed Flight Officer Darnaby H. Wilhoit, United States Army Air Force

On 29 May 1945 23 Mustangs earmarked for transhipping to the Japanese theatre of operations took off from Debden, in Essex, to be ferried to Speke, Liverpool. Met reported that low cloud was shrouding all high ground, but though the route was to pass over the Peak District, the pilots were briefed that the en-route terrain was 'flat and level'.

At the appropriate time Flight Officer Darnaby Wilhoit, in Mustang 44-64084, duly followed his section leader down to 1,500 feet. At that level, however, the section leader, finding himself still in cloud, powered up and commenced a climb, possibly rather too hastily, for in doing so he lost his three followers. Of these, Flight Officer Wilhoit attempted to fix himself, his aircraft being heard to circle the vicinity of the Grouse Inn, Glossop. It was then seen to break cloud at a low altitude, make a hasty terrain-avoiding turn, then disintegrate as it struck the ground. Flight Officer Wilhoit's watch, stopping at 1054 hours, furnished poignant proof that his final flight had lasted just 49 minutes.

Of the other Mustangs, the formation leader, letting down blind, crashed 35 miles to the north-east of Speke, while his formating number two clipped a wing on the ground but was able to climb away and return to Debden. The rest, in pairs or as singletons, either probed their various hazardous ways into Speke, or aborted the sortie.

The site, from which the wreckage was cleared within a few days, is on private farmland but accessible from Monk's Road. In 2009, although no surface evidence remained, a confirmatory metal-detector sweep showed scraps of debris fanning out from the impact site, just beyond the field's gate.

Bell 206B Jet Ranger Helicopter G-ODIL

William Clough, Kinder

SK 06655 89697	533 m
Owner	Yorkshire Helicopters
Date	24 October 1997
Crew	Pilot (name not recorded), uninjured

Jet Ranger G-ODIL was engaged in lifting stone from the shoulder of the Ashop Head Scarp, at the summit of Kinder's William Clough, when it suffered a dynamic-rollover upset: that is, it tipped over. The aircraft was later airlifted to the A57 at Doctor's Gate for disposal.

The site is reached by turning hard right (170°M) from the foot of Ashop Head Scarp and walking off-path for 200 yards. In accordance with good civilian practice, nothing of any significance was left at the site, although a few items of debris were held at the Hayfield Ranger's Station near Bowden Bridge.

8

Miles Hawk (RAF Magister) G-AJSF

Kinderlow End, Kinder Scout

SK 07368 86688	594 m
Operator	Blackpool Aero Club
Date	29 July 1957
Crew	Pilot, killed Mr William Warburton Hall

The former-RAF Magister trainer, Miles Hawk G-AJSF of the Blackpool Aero Club, crashed fatally at Kinderlow End on 29 July 1957. The pilot, Mr William Hall, an ex-RAF wartime pilot, took off from Squires Gate, Blackpool, for the 34-mile flight to Manchester's Barton airfield, envisaging being airborne for just over 15 minutes. Mr Hall did not consult the meteorological service but he could hardly have been unaware of the pattern of strong winds. Some time later walkers saw the Hawk pull into a steep turn which enabled it to avoid the Kinder Scout summit but took it into the cloud-enveloped Kinderlow End. The inquiry decided that, flying above cloud, Mr Hall had sped past his destination.

The site is virtually on the ridge footpath leading to Kinderlow End. Despite the remoteness of the crash site, the Hawk was easy to salvage, so that although in 2003 a tiny scrap of yellow canvas was discovered, by 2010 no trace could be found.

Lockheed Hudson (spurious)

Edale Moor

SK 10100 87800	610 m

Enthusiast lists hold that a Lockheed Hudson came down just '50 yards from Anson N9853 on Edale Moor', and that the RAF 'dragged the wreckage away over the snow'. As late as 2010, however, and despite intensive inquiries, nothing had been found to substantiate this. The report is included here merely for completeness.

10

Cessna 150M G-BFRP

Broadlee-Bank Tor, Edale

SK 11390 86120	490 m
Operator	BTJ Aviation Group (Transgap Limited), Ringway, Manchester
Date	23 October 1983
Crew	Instructor and student, both survived Mr Barry Bryant, pilot, instructor Mr John Steward Bateson, student pilot

Reims Cessna 150M G-BFRP was being homed under radar advisory service towards Manchester after a night training flight when it was caught in a lee-wave downdraft and crashed into the mid-slopes of Broadlee-Bank Tor, above Edale village. Initially it struck near the top of a steep gully, then toppled backwards to come to rest inverted. The aircraft was destroyed but although both Mr Barry Bryant, the instructor, and Mr Steward Bateson, his student, suffered bruising on letting themselves fall from their seatbelts, they were able to walk down to Edale, push their way into the gathering rescue party, and report the accident.

The subsequent investigation found that Mr Bryant, a private pilot with over 2,500 hours' flying experience, had accepted radar assistance on condition that he did not exceed 3,000 feet and that he remained responsible for terrain clearance. When he declared an emergency he was dutifully at 3,000 feet but in the lee (downwind) of the 2,088 feet Kinder plateau where a 52 mph (45 knot) wind was creating severe downdraughts. Moments after Mr Bryant radioed that he was 'going down at a thousand feet per minute', the aircraft struck the ground at 1,640 feet above sea level.

The next day a team including Mr Gordon Miller (at that time Area Ranger for Kinder – including Edale, Castleton, and Hayfield) cleared the wreckage in order to prevent spectacle-seekers damaging the drystone wall between the upper moorland and the lower pastures.

The site is best accessed by taking the shooters' path leading from Edale School towards Grindslow Knoll, then branching off south-westwards at SK 11630 86350. On reaching the ruined barn the site is directly upslope, however, the steep-sloping mix of grass, bracken and bilberry of the hillside shows no trace of the accident.

11
Unidentified aircraft

Longdendale Valley

Date	c.1935

In June 2006 researcher Mr Arnold Willerton, of Hyde, supplied a copy of a photograph from *The Longdendale Valley*, by Mrs Margaret Buxton-Doyle, of Denton. It was taken by Mrs Buxton's father, Mr Harry Buxton (1908-83), and shows an aircraft which crashed in the valley in the 1930s. The incident is recorded here in the hope that more information may be forthcoming.

12
De Havilland DH9A

Rollick Stones, above Torside Reservoir

SK 08137 98602	359 m
Unit and Station	RAF, on delivery from No. 2 Northern Aircraft Repair Depot, Coal Aston, also known as Norton
Date	17 October 1919
Crew	Two, both injured Lieutenant A.S.M. Meydrick-Jones, RAF (the most likely identity) Observer, presently unidentified

This crash occurred on the rocky shoulder below Rollick Stones, to the west of Fair Vage Clough, and high above the junction of the Torside and Woodhead Reservoirs.

Unsubstantiated references hold that this DH9A was on a delivery flight to Liverpool from No. 2 Northern Aircraft Repair Depot at Norton (Coal Aston) near Sheffield, when it ran into low cloud. An article from the *Glossop Chronicle* of 24 October 1919, (supplied by Ranger Phillip Shaw, of Glossop), records that while one of the occupants was only slightly hurt, the other was taken to a Sheffield hospital. In a poignant addition it further records that the crash was witnessed by Mrs Sarah Ann Shaw (38), of Railway Cottages, Woodhead, who set out with the intention of lending aid but suffered what was thought to have been a heart attack, and died, leaving a husband, a son, and three daughters.

In 1970, the site location was supplied to researchers Mr John Ownsworth and Mr Jim Chatterton by Mr John Davies, the then-incumbent of Railway Cottages, Woodhead. At that time they found an abundance of wooden fragments but also a brass petrol cap stamped by the Aeronautical Inspectorate Directorate. More recent searches, including one made in company with researcher Mr Alan Jones, found no trace whatsoever.

A satisfactory way to visit the site – which, if nothing else, gives spectacular views of the Longdendale Valley – is to park near Railway Cottages (SK 08245 99360) and follow the barbed-wire boundary of the Boar Pigeon Shoot. The most straightforward way would be to obtain prior permission from the operator of the shoot to utilise the range's access tracks.

13

Heinkel He111

Near Lady Cross, Langsett Moor

SK 14811 99586	461 m
Luftwaffe	
Date	possibly 17 November 1940

A Heinkel He111 ('One eleven') is widely believed to have come down on the Langsett Moors close to Lady Cross. Other accounts have the raider coming down in 'the moors behind the Dog and Partridge public house'; the pub being at the bottom of Bordhill on the southern side of the A628. This was the version given to researcher Mr John Ownsworth by Mr Stephen Marsden, whose father, a butcher at Thurlstone, remembered that the aircraft had been trailing smoke from one of its engines prior to crashing. Another witness vouching for the smoking engine specifically held that the crash occurred on 17 November 1940 and that the machine had been coming from the direction of Manchester at very low level.

Lady Cross is most easily reached from the footpath leading off the A628(T) at Lasche (SE 15550 00050), but parking is to be found rather further west, at SE 13600 00040. Having reached Lady Cross, a minor path leads southwards to pass close to the site which is in unbroken heather, some 200 yards from the Cross. As might be expected, nothing is to be seen.

14

Miles Magister Mk.1 N5418

Upper House Farm, Cowms Moor, Woodlands Valley

SK 12499 90076	401 m

Unit and Station	No. 145 Squadron, RAF Catterick, No. 11 Group, Fighter Command
Date	28 July 1941
Crew	Pilot, uninjured Sergeant Pilot W.J. Johnson

An RAF accident investigation summary records that on 28 July 1941, in the course of Spitfire-equipped No. 145 Squadron's move from RAF Merston, near Chichester (Sussex), to RAF Catterick, in Yorkshire, Sergeant W.J. Johnson was detailed to ferry the squadron's Magister runabout. While doing so he encountered bad weather in an area of high ground and elected to carry out a precautionary landing. The touchdown was made on Cowms Moor in the Woodlands Valley, in a field belonging to Upper House Farm, but the aircraft was damaged when a wheel ran into a hole on rough ground. Sergeant Pilot Johnson was not held to blame, however, and no disciplinary action was taken.

A footpath following the line of the Roman Road runs from Hayridge Farm towards Upper House Farm. Reaching the site of this set-down would require a 300 yard excursion northwards from the path towards Cowms Rocks.

15
German Bomber

Crook Hill area, Woodlands Valley

SK 18281 86695	315 m (area of)
Luftwaffe	
Date	1940-42

Mr George Hallam, who served at the joint Home Guard and Observation Corps post on Crook Hill throughout the war, left a description of this event. 'The plane came down,' he wrote, 'on the flat moor on the far side of the hill from Crookhill Farm. The three crew members gave no trouble, but as the post's communications had broken down they had to be escorted to Stephen Elliot's farm [Crookhill], which had a telephone. The responsible officer, billeted at the Derwent Hotel, in Bamford, directed that the airmen should be held until next day when Regular Army personnel could collect them. So us Home Guard watched the prisoners, two hours on, two hours off, while the Elliot family watched everyone.'

After a courtesy call at Crookhill Farm, any site visit would require an undemanding off-track walk to the far side of the hill. No sign of the incident remains, nor have local enquiries borne any fruit.

16

Junkers Ju88

Low Tor, Howden Edge, Howden Moors, east (inconclusive)

SK 20196 92209	466 m (area of)
Luftwaffe	
Date	c.1941

Enthusiast lists claim that a Junkers Ju88 came down 'near the shooting cabin to the north of Low Tor', on Howden Edge, and – with more substance – that German ammunition was found at the reference given above. This proves to be a marshy patch in Bents Clough, to the north of Low Tor, and high above Abbey Brook.

By 2010 nothing more had been discovered, except to establish that the shooting cabin, like so many others, has long been demolished. The reference, then, can best serve as a lure to entice walkers from where limited parking is available at Strines Bridge (SK 22100 90900) up to Howden Edge and the Cartledge Stones Ridge Path, for though the marsh might be found devoid of interest, taking just a few score extra paces towards Berristers Tor – to overcome the convexity of the slope – affords a spectacular view over Abbey Brook.

17
Junkers Ju88

Luftwaffe	KG106 (*Kampfgeschwader*: bomber group)
Date	3 July 1942

The details of this aircraft crash were recorded by *Dark Peak Aircraft Wrecks* author Ron Collier, who also named crew members Bergman and Majer. The entry is recorded here in the hopes that more information might be forthcoming.

18
De Havilland DH90 Dragonfly

Thornseat Delf, Bradfield Moors

SK 22972 92717	429 m
Operator	RAF. Civil aircraft impressed to Service duty
Date	c.1941

The five-seater, de Havilland DH90 Dragonfly cabin biplane, the luxury version of the twin-engined de Havilland Dragon Rapide, was one of many civil types to be impressed into RAF service during the Second World War. The location where one of them crashed at Thornseat Delf [drain] was pointed out by Mr Alwyn Haigh of West Nab Farm, Bradfield. 'I would have been about twelve,' he remembered, 'which would make it about 1941. I'd no idea what type it was but we could clearly see its tail sticking up from the top of the bank [hillside] just as if it had nose-dived in.'

The area may be approached by the Thornseat Road track which leaves Mortimer Road at SK 23600 91910 where there is very limited lay-by parking. The site indicated by Mr Haigh is on the brow of the ridge just above the two shooting cabins in Small Dale (Emlin Dike), but unsighted from them by the convex slope.

19

De Havilland Tiger Moth DH82A EM931

[Thornseat Delf], Bradfield

SK 26878 93215	329 m (RAF Form 1180 gives contemporary MR sheet 37 735 135), half a mile north of Bradfield
Unit and Station	No. 25 (P) EFTS, RAF Hucknall, No. 21 [No. 21] Group, Flying Training Command
Date	9 July 1945
Crew	Pilot, uninjured Pilot-under-Training Aircraftman Two Stanisław Wójcikowski, Polish Air Force under British Command

Pilot-under-Training Stanisław Wójcikowski had done 25 hours of solo flying when he was detailed for a training sortie based around aerobatics. In the course of this he became lost and after being airborne for just an hour and 25 minutes, feeling that he was running short of fuel, put down in a field. The surface of the field proved to be unsuitable and the aircraft was damaged. The court of inquiry felt that the pupil pilot had been unduly fearful of his engine stopping for lack of fuel. The Air Officer Commanding then supported the recommendation that Pilot-under-Training Wójcikowski's record should be endorsed 'Loss of good conduct' and that he should be 'Confined to Camp' for five days (CC, or 'jankers', the most minor form of detention), the AOC in Chief concurring.

It is probable that this incident, long taken to relate to the c.1941 Dragonfly crash (see above), was the one attended by former-corporal Charles D. 'Taffy' Austin, BEM, of the Harpur Hill Mountain Rescue Team, and reported to Mr John Ownsworth in the 1970s. This follows from the fact that the team only really came into being in 1942. Then again, Mr Austin reported a tail sticking up from 'a marsh'; a reference which puzzles locals. Equally puzzling is that the crash-report map reference gives a location just north of Bradfield while its written location, of Thornseat Delf, is over two miles to the south-west. True, on occasion RAF teams did get local names for places mixed up. On the other hand, Mr Alwyn Haigh, of West Nab Farm, Bradfield, a fount of local knowledge, could recall no aircraft coming down nearer Bradfield than Thornseat Delf. A fine case, then, for future research.

20
German bomber (spurious)

Agden Bridge, Broomhead

SK 24293 93794	307 m

Although a German bomber crash at this reference has a long provenance in enthusiast lists, investigations have found no substance in the reports. The almost certainly spurious site is included here merely for completeness. See too, the similar entry regarding Ewden Bridge (below).

21
German bomber (spurious)

Bull Clough, Ewden Bridge, Broomhead

SK 23705 96790	223 m
Date	1940-41

That a German bomber crashed at this site has anecdotal provenance lodged in otherwise dependable sources dating back to 1946. Suffice to say that by 2010 nothing of substance had emerged. As is the way with popular fancy, however, RAF aircrew seeking aid after a local Stirling crash (see above) passed into lore as German bomber crews, similarly nearby wreckage was held to be adorned with swastikas. Certainly, Broomhead Estate keepers David Beaumont and Chris Cunningham, familiar with the Moor for many years, knew nothing of wreckage nearer Ewden Bridge than what was provenly from the Stirling. The supposed incident, then, is recorded here for completeness.

Ewden Coppice, from Ewden Bridge to Bull Clough, though for the most part sadly overgrown with bracken and brambles, makes a delightful short walk. There is very limited parking (*clear of the drive!*) at the top of the slope at SK 24250 94890.

22

German aircraft (spurious)

South-east of Emlin trig column

SK 24201 93294	364 m

This spurious record, which appears on enthusiast lists, arose from a misunderstanding in the 1970s when Mr Alwyn Haigh of West Nab Farm, Bradfield, meaning to show Mr John Ownsworth the Dragonfly site on Thornseat Delf – a location just three degrees to the left of the Emlin reference above – pointed with outstretched arm: a misunderstanding subsequently conceded by miscreant Mr Ownsworth. This entry, therefore, aims to save other walkers investigating the enthusiast list from an unremitting, unrewarding wallow through thick, progress-resisting heather.

23

Avro Anson Mk.1 N9912

Whitwell Moor, Stocksbridge

SK 24758 97491	345 m
Unit and Station	No. 25 Operational Training Unit, RAF Finningley, No. 7 Group, Bomber Command
Date	31 March 1941
Crew	Four, superficial injuries Pilot Officer Bernard Maurice Fournier, RAF Volunteer Reserve, pilot Sergeant Duncan Henry Barrett, navigator Sergeant Ernest Richard Palmer, wireless operator/air gunner Sergeant Dennis Watson, wireless operator/air gunner

On 31 March 1941, flying Anson N9912 on a night navigational exercise, Pilot Officer Bernard Fournier's trainee crew, unable to make wireless contact, became uncertain of their position. Pilot Officer Fournier chose to descend through cloud to obtain a pinpoint, but bellied into a gently sloping hillside. Nobody was seriously injured and the court of inquiry put the accident down to inexperience, observing that the crew member actually manning the wireless station had spent most of his productive flying in his dual-category role as an air gunner, which meant that he had lost the necessary edge called for by the somewhat temperamental sets of the day.

The crew were subsequently posted to No.49 Squadron on Handley Page Hampden bombers at Scampton only to be shot down in flames by a night-fighter on 29 August 1941, having raided Duisburg.

The Whitwell Moor site, a seasonal pool on a heathery rock-strewn slope, bears no trace of a crash. In 1981, however, debris enabled researcher Mr John Ownsworth to re-identify the site. It might be difficult to decide upon the right pool – in early 2010 it was the highest and longest of three pools, marked by its flourishing rushes.

24

Trainer, Elementary Flying Training School type

Moor House Farm, Long Lane, Whitwell Moor, near Stocksbridge

SK 24862 97971	312 m
Date	1939-45
Occupant(s)	Unidentified, uninjured

During the Second World War the pilot of a light training aircraft, lost and short of fuel, put his aircraft down, sensibly choosing this pasture to the north of Long Lane, Whitwell Moor, and close to Moor House Farm. The aircraft remained on the ground for some time but once it was refuelled, the incumbent of the farm, Mr Nelson Kay, helped turn it around, after which it took off again. Unsurprisingly, nothing more is known of this successful precautionary landing, the incident being recorded here mainly because the field is overlooked by the prime walking area which holds the no-debris crash site of Anson N9912 (see above).

25

Lockheed Hudson AM531

Deepcar, Soughley Bridge, north-east of Stocksbridge

SK 29103 98274	146 m, touchdown area
SK 29228 98285	138 m, terminal area in ravine
Unit and Station	No. 6 (Coastal) Operational Training Unit (OTU), RAF Thornaby, No. 17 Group, Coastal Command
Date	4 November 1942
Crew	Three, all suffered minor injuries Flying Officer William Hampson, OTU staff pilot Pilot Officer G.M.I. Tweedie, staff navigator/bomb aimer Flight Lieutenant Leslie Roy Aust, wireless operator/air gunner

In the early evening of 4 November 1942, Flying Officer William Hampson, flying Hudson AM531 in conditions of low cloud and poor visibility, and unable to make voice-radio contact, decided to make a precautionary landing on one of the region's enormously extensive slag heaps, in this case nearly a mile long and half a mile wide! In accordance with contemporary practice he made a wheels-up approach, but after touchdown the aircraft hit an obstruction and slewed off into a deep gully. The crew suffered only minor injuries although the Hudson had to be scrapped. Significantly, the court of inquiry laid some blame on the Signals staff at St Eval, the departure airfield, for failing to supply the crew with the correct frequencies-for-the-day crystals; a rare occurrence of failure by ground personnel!

Hudson AM531 was soon salvaged, after which restructuring of the site for the major A616/A6102 junction concealed any debris that might have remained. Just the same, the sparkling River Porter – the little Don – which threads the now-sylvan slopes of the former slag heap has justifiably made it a prime venue for local walkers.

26
Vickers Armstrong Wellington TMk.10 MF627

Rod Moor, Ughill

SK 26396 89270	351 m
Unit and Station	No. 6 Air Navigation School, RAF Lichfield, No. 21 Group, Flying Training Command
Date	22 October 1952
Crew	Three, two injured Sergeant Reginald A. Keith, pilot Pilot Officer David E. Ward, navigator under training, injured Pilot Officer Brian Thirkell, navigator under training, slightly injured

On 22 October 1952 Sergeant Reginald Keith was detailed to fly two trainee navigators on a night-navigation sortie. The exercise completed, the operating trainee, Pilot Officer Ward, was called upon to set up for a crew-controlled let-down using the Gee radar equipment (see Glossary). Suffering from air sickness, however, he made a mistake which showed the aircraft to be considerably further south than it actually was. When Sergeant Keith began to lose height, therefore, still in darkness and with cloud below him, he was unaware that he was, in fact, over high terrain and some 52 miles north of base. In the event, both he and the off-watch trainee, Pilot Officer Thirkell, saw the loom of rising ground at the same moment, but Sergeant Keith had no time to do any more than pull up the nose before the aircraft struck.

Having smashed through a section of drystone wall, MF627 slewed about before coming to rest pointing downslope but, thanks to the ruggedness of its construction, with the forward crew area still intact. For reasons unstated Pilot Officer Ward was obliged to make a brief visit to hospital ... Later he ceased training.

Parking is available on Corker Road (south of Ughill) at SK 26440 89800. Following the south-easterly farm-cum-quarry track leads to a junction at SK 26523 89574. The site then lies 360 yards off on a heading of 199°M. Although minor debris was unearthed, essentially, apart from a gap in the drystone wall, reopened by weather long after the RAF had rebuilt it, there is nothing to see but gently rolling pastures that stand in quiet contrast to the rugged heather moorlands to the west.

27

Hawker Hart K4423

Peacock Lodge Farm, Wentworth

SK 40000 96700	55 m
Unit and Station	No. 9 Flying Training School, RAF Thornaby, No. 23 (Training) Group, Training Command
Date	23 July 1937
Crew	Pilot, uninjured Flight Sergeant Raymond Middleton

On 23 July 1937 Flight Sergeant Raymond Middleton ran into bad weather and made a precautionary landing near Peacock Lodge in Wentworth Park, Greasbrough. He touched down successfully but then ran into a fence, damaging both wings. No RAF crash report exists and a *Sheffield Telegraph* report gave no specific location. The fields below the farmhouse, however, seem eminently well suited to making a planned set-down.

28
Supermarine Spitfire K9941

Back Lane, Hooton Roberts, Rotherham

SK 48601 96945	46 m
Unit and Station	No. 72 (Basutoland) Squadron, RAF Church Fenton, No.12 Group, Fighter Command
Date	21 August 1939
Crew	Pilot, killed Sergeant Donald Victor Peacock

On 21 August 1939 Sergeant Donald Peacock, of No. 72 Squadron, Church Fenton, Yorkshire, was engaged in local flying when his engine began to malfunction. He lowered the nose and commenced a forced-landing. Late on the approach, however, his engine picked up again. He lifted the nose, only to have the engine cut once more. Caught nose high, he then stalled and fatally crashed into a field off Back Lane at Hooton Roberts, near Rotherham. There was a fire after impact, but the coroner determined that Sergeant Peacock had been killed instantly when his head struck the instrument panel.

Mr John Wheelhouse of Holly Farm, was able to verify the actual impact site. 'Some 25 years ago,' he explained, 'a metal-detectorist got permission from the then-owner, Mr Beaver, to dig in the field. I had a JCB handy, and we found several bits of aircraft metal.' Certainly, in early 2010 there was no sign of the crash in what had since become known as Beaver's Top Field.

29

Handley Page Halifax W7815

Bassingthorpe Farm, Greasbrough, near Rotherham

SK 41764 94329	56 m
Unit and Station	No.1652 Heavy Conversion Unit, RAF Marston Moor
Date	9 December 1943
Crew	Seven, two injured Flight Lieutenant David Roy Fisher, DFC, DFM, pilot, injured Sergeant Stephen Thompson Wells, pilot, injured Sergeant James Royden Weaver, navigator Sergeant Reginald Henry, flight engineer Warrant Officer Allen, wireless operator/air gunner Warrant Officer Hughes, wireless operator/air gunner Flight Sergeant Gray, wireless operator/air gunner

This heavy bomber got into difficulties during a conversion sortie from RAF Marston Moor and was set down on farmland at Bassingthorpe Farm, Greasbrough, near Rotherham. It ploughed its way towards a farmhouse but stopped some way short on impacting with a wall bisecting a duck pond.

This incident had been attributed to 'an unidentified trainer' but Mr R. Peter Fisher advised that the aircraft concerned was in fact, Halifax W7815. Mr Fisher's parents, living at Brown Riddings, a quarter of a mile off, and having put up two of the recovery crew, remembered that the ruts in the field showed how the pilot had been forced to struggle to keep the aircraft straight.

There is nothing to be seen at the crash site, but Mr Alan Wood, the incumbent of the farm, identified the location of the former pond.

30

Hawker Hind

Higher Haugh, near Greasbrough, Rotherham

SK 42057 97484	109 m
Unit and Station	Royal Auxiliary Air Force
Date	18 January 1939

On 19 January 1939 the *Sheffield Star* carried the report of a Hawker Hind being forced down by bad weather at Stubbin, Higher Haugh, near Greasbrough, on land owned by farmer Mr Hobson. The field was adequately sized and level enough, but having been newly sown, was soft, so that although the aircraft was undamaged it could not be safely flown off. In early 2010 the then-incumbent knew nothing of the incident. The location given, therefore, is the most likely one of the original Stubbin fields to be chosen for a precautionary landing.

31

Handley Page Hampden Mk.1 P1248

Concord Park, Shiregreen, Sheffield

SK 38025 92781	100 m, impact site
SK 37391 92002	144 m, where the pilot landed, subsequently a sports centre
Unit and Station	No. 25 Operational Training Unit, RAF Finningley, No. 7 Group, Bomber Command
Date	19 April 1941

Crew	Two, one killed Pilot Officer Ralph Athelsie Pole Allsebrook, pilot Pilot Officer Jeffery Bohun Ranson, observer, killed

On 19 April 1941, Pilot Officer Ralph Athelsie Pole Allsebrook, a pupil at No. 25 Operational Training Unit, was briefed to carry out an instrument-flying practice in Hampden P1248. The exercise required the pilot to fly eyes-down on his instruments while carrying out various flight manoeuvres, the lookout being maintained by his observer – navigator – Pilot Officer Jeffery Bohun Ranson. By the flight commander's decision, the wireless operator was stood down.

As it was, Pilot Officer Allsebrook strayed from the local area and penetrated the Sheffield air defences where, failing to hear, or to be advised of, the automatic warning signals sent out by the barrage balloons, he flew into a tethering cable at 2,800 feet. The aircraft then crashed into a woodland fringe in Concord Park, on the north-eastern outskirts of Sheffield, catching fire and burning out. Pilot Officer Ranson had been killed by the cable but Pilot Officer Allsebrook, although injured, parachuted to safety.

The court of inquiry found that the crew had lost themselves and that 'the observer, who was also the "look-out", failed to keep check of the pilot's course on the map.' The Air Officer Commanding, for his part, further observed that the root cause of the fatal crash was the lack of explicit orders for the sortie. This, he put down to 'the inexperience of the instructional staff', observing further, 'a common failing in OTUs.' He then made the specific charge that no wireless operator had been carried to monitor warning signals and obtain bearings, scathingly attributing this to 'a misconception on the part of the flight commander', and ruled that in future, wireless operators would be carried on all flights.

Mr Terence Crooks, a long-term resident of the area, described where the cable-stricken aircraft crashed. 'It was just inside Woolley Wood, not far from the Concord Road entrance. Or from where they later built some brick toilets. I was only about eight at the time, but we all wanted to get souvenirs. But the police and the RAF chased us off. We wanted souvenirs, whereas all we got for our pains were thick ears.' There is nothing to mark the actual impact point, while the place where Pilot Officer Allsebrook had parachuted down had long been built over by the Concord Sports Centre. Both locations were vouched for, however, by Mr Patrick Flynn, of Preston, who knew the Park well.

32
Handley Page Hampden

Concord Park, Shiregreen, Sheffield

SK 37606 92368	132 m
Unit and Station	RAF Bomber Command
Date	8 July 1943
Crew	Probably four, uninjured

Sheffield's Concord Park, in the Shiregreen area, to the north-east of the city, is a generously open public space utilised by a golf course, various sports fields, a children's playground, and woodland walks. In 1943, however, it was the scene of a set-down by a Hampden bomber. Whether this was a forced-landing or a precautionary landing is not known. Indeed, little enough seems to be recorded about this incident.

Mrs Maureen Bailey (née Rodgers), however, encountered dog-walking, knew something of it. 'My older sisters, Sheila and Margaret,' she remembered, 'have often told me how they went to see it, and showed me where it happened – I was just a baby, of course. It had come down in what was then a cornfield not that far from the golf course. But it wasn't damaged and was flown off again just days later.'

This must remain another site, then, to be recorded for posterity. Just the same, Mr Patrick Flynn, of Preston, a gardener's mate in Concord Park from late 1942 to 1944, was able to indicate the area – still a wide space in 2010 despite the encroachment of the golf driving range in the 1990s – where both this Hampden, and a Tiger Moth (see below), successfully put down.

33

De Havilland DH82A Tiger Moth

Concord Park, Shiregreen, Sheffield

SK 37543 92235	130 m
Unit and Station	RAF Training Command
Date	Late 1942-1944
Crew	Solo pilot, unidentified, unhurt

Mr Patrick Flynn (see above) remembered standing guard on a Tiger Moth while the pupil pilot, who had got lost and run short of fuel, went to find a telephone. Eventually an RAF support crew arrived, but in driving their truck directly across the park and proceeding to refuel the trainer with petrol from jerry cans – leading, no doubt, to a certain amount of spillage – they incurred the displeasure of Mr Flynn's boss, the head gardener. Indeed, the impression given is that only after placating him was the aircraft allowed to be flown off to its home station.

34

Handley Page Hampden Mk.1 L6011

Hellaby, Maltby

SK 49723 92332	104 m (area of)
Unit and Station	No. 5 Air Observers' School, RAF Jurby, No. 25 Group, Flying Training Command
Date	26 May 1942

Crew	Five, one killed, four injured Sergeant J.W.G. Hodgkinson, pilot, seriously injured Other crew members presently unidentified

In the course of a night cross-country training flight, the pilot's radio failed, as too did his artificial horizon, his main attitude reference. At about midnight, therefore, when a throttle linkage fractured, leaving him with no control over the starboard engine, he was only too grateful to see what appeared to be the flarepath of a night-operating airfield. Low on the approach, however, he hit electrical power cables, the aircraft catching fire and crashing. One crew member was killed, and the rest injured, Sergeant Hodgkinson severely.

Tragically, what Sergeant Hodgkinson had seen was a decoy airfield – a 'Q' Site – in this instance a subsidiary part of a whole complex of lights and fires intended to invite German bombers to an area of open ground at a safe distance from nearby built-up areas and operational airfields.

The subsequent enquiry absolved Sergeant Hodgkinson of blame, but the Air Officer Commanding in Chief did insist upon knowing why none of the Q-site's personnel – normally up to half a dozen men – had been at hand to switch off the lighting when it was seen that an aircraft was actually making an approach. The result of this follow-up is not known.

Retired fireman Mr Derrick Waters, of Doncaster, remembered, 'As a lad of six, I was living at Newhall Grange. I saw this aircraft from my bedroom window, clearly on fire. I called my mum, who hurried upstairs just in time to see it. Initially it was heading towards us, but then it began a turn which would have taken it to what is now the Hellaby Industrial Estate, on the far side of the M18. At that time this was open ground and full of inflammable rubbish which they set alight to decoy German bombers away from Sheffield. I then saw a great flash as the aircraft hit power cables and crashed into the field.'

Nothing is to be seen at the site, which is to the west of and parallel to Junction One of the M18.

35

Vickers Wellington

Lilly Hall Farm, Maltby

SK 51315 93051	127 m
Date	1939-45. No other details available

Mr Derrick Waters, of Doncaster, remembered that a Wellington was said to have crashed at Lilly Hall Farm, Maltby. The parachuting crew, as he heard the story, landed in the Maltby Dike area, parallel to Rotherham Road, and hid, fearing they were in enemy-occupied territory, until they saw an Oxo van. Mrs Josephine Fragola was brought up on Lilly Hall Farm and her husband, Joseph, was able to point out the area, near a small copse, where the aircraft came down. Family tradition has it that the pilot was brought to the house, but died.

The area of the site is best accessed from Lilly Hall Road, Maltby. There is access by a long track running from north-west of Maltby, but this is not recommended for cars, especially in wet weather, regardless of the Satnav's persuasion.

36

Zenair Zodiac HC601 G-YOXI

Hellaby Park Farm, Ravenfield, Rotherham

SK 50350 93598	103 m
Owner	privately-owned, 'home-build' kit aircraft
Date	25 August 2006
Occupants	Both killed Mr Brian Leslie Yoxall, owner, and operating pilot Mr Terence John Whitfield, passenger

At 1630 hours on 25 August 2006, a privately-owned, 'home-build' kit aircraft, Zenair Zodiac HC601 G-YOXI, suffered a catastrophic structural failure while essaying to carry out a low-level pass at Hellaby Park Farm, near Rotherham, crashing and burning and killing both occupants. The aircraft was owned by Mr Brian Yoxall, a builder, who had just 200 hours of flying experience, all on very light aeroplanes. The passenger, farmer Mr Terence Whitfield, had been for many years a civilian ground instructor with No. 300 Squadron of the Air Training Corps at Crowle.

The two friends had departed from Mr Whitfield's landing strip at Askern, to the north of Doncaster, with the intention of carrying out a light-hearted aerial inspection of three newly-seeded fields at Hellaby Park Farm, a landing strip shown on the aeronautical chart as Ravenfield Helipad. Prior to getting airborne they had telephoned for permission for the overflight, had been passed a contact frequency, and advised of the 30 foot high power cables crossing not that far from one end of the strip.

Speaking just days after the accident, Mr Ray Wharam, the owner of Hellaby Park Farm, described the crash. 'The aircraft touched the ground,' he said, 'bounced, then struck hard some yards further on, spinning about.' Standing once more at the crash site, he pointed to his original tyre tracks, curving in to within feet of the soiled patch of fragment-strewn earth. 'There were already flames licking up, but for a moment, as I reached the cockpit and looked in, I thought my fire extinguisher could deal with them. But suddenly the whole thing just erupted; and everything began to dissolve in front of me ...'

He took a moment then, to collect himself. 'I drove back to the house and hitched up my water bowser. By the time I got back to the aircraft, however, a policeman had come down from the motorway, and wouldn't let me near.' Again he paused, but then said soberly. 'Not that there was anything to be done.' He reflected. 'The only heartening thing was that every motorist I saw come down the banking had a fire extinguisher: clearly they'd come to help, not simply to rubberneck.'

And what had Mr Wharam been about, leaning over a burning cockpit as he had done? He smiled grimly. 'I've done fire courses. And I know how I'd be criticized. But when it comes to a question of human life ...' And he pragmatically brushed aside all notion that his action had been in any way heroic.

The hard-packed ground has left nothing to be seen, nor is there a monument, beyond which the site is remote from any public footpath and on strictly private land.

The tragedy forms a saddening coda to this book, in which the other accidents recounted are relatively well distanced by comforting Time; reflecting the agelessness of Captain Lamplugh's celebrated 1931 observation: '*the air, to an even greater extent than the sea, is terribly unforgiving [and] carelessness, neglect or overconfidence are paid for more quickly and more dearly than in other forms of transport*'. The Civil Aviation Authority's (CAA) accident report was duly published in May 2007. It found that the harsh pull up near the farm-strip cables had caused the main spar to fail, further recording that two months earlier the pilot had been the recipient of a CAA reproof for making steep-dive low passes, followed by steep turns.

So Captain Lamplugh's words remain prescient! How 'terribly unforgiving', indeed, the air continues to be to overconfidence and how quickly and dearly it can exact its payment!

Aircraft Types

This section aims to provide the moorland walker-reader with a (very) potted guide to the once-proud aircraft mentioned in this book, now represented, at best, by mere shards of debris. Arabic numbers are employed throughout in designating aircraft marks, the problem of deciphering Roman numerals being left to the enthusiast. As for performance figures, published sources dealing with wartime aircraft often perpetuate values enhanced for propaganda purposes. But then even those quoted in *Pilot's Notes* incorporate a healthy safety margin, while workaday machines of a given type differ significantly.

Airspeed Consul The Consul was the post-war passenger version of the RAF's twin-engined Oxford. It could carry six passengers and two crew and was popular with emergent air operators.

Powered by two 375 horsepower Armstrong Siddeley Cheetah Ten engines, it had a maximum weight of 8,250 pounds, cruised at 163 mph (142 knots), had a ceiling of 19,000 feet (with a variant claim of 23,500 feet), and a range of 900 miles.

Airspeed Oxford The twin-engined, wooden-framed, plywood-skinned Airspeed Oxford, from which the Consul (see above) derived, first flew in June 1937 and remained in RAF Service until 1954, almost 8,600 being built. This dual-controlled, general-purpose trainer had a basic crew of three but could accommodate other trainee aircrew depending upon the instructional role in hand.

Two 375 horsepower Armstrong Siddeley Cheetah Ten radial engines, or alternatively, two 450 horsepower Pratt & Whitney radial engines, gave it a cruising speed of some 163 mph (142 knots) and a ceiling of 19,000 feet. It had a maximum take-off weight of 8,000 pounds, and a range of 700 miles. It could carry practice bombs and a few had a dorsal turret (see Glossary) with a single 0.303 inch (7.7 mm) calibre machine gun.

Avro Anson (see illustration on page 39) The 1935 Anson stayed with the RAF for 22 years. Originally conceived as a maritime reconnaissance aircraft, it was withdrawn from operational service in early 1942, after which it was employed extensively in the training role. As a dual-controlled machine with such innovations as hydraulically operated flaps and undercarriage, it was used to train most aircrew specialisations. Normally accommodating between three and five, it could also be configured as an eight- to eleven-seat communications aircraft. The Anson was well liked, being easy to fly, dependable, sturdy, and relatively forgiving. Its performance on one engine, however, was poor.

With a maximum take-off weight of 8,000 pounds, the Anson was typically powered by two 350 horsepower Armstrong Siddeley Cheetah Nine radial engines. These gave it a cruising speed of 158 mph (138 knots), a ceiling of 19,000 feet, and a range of nearly 800 miles. In all, some 11,000 were built, many in Canada.

Avro Lancaster The Lancaster, a four-engined machine which metamorphosed from the twin-engined Manchester, became operational in 1942. Designed for ease of production and subsequent servicing, 7,737 were built by 1946. Several constructors were employed, among them the Victory Aircraft Company of Canada which produced 430 Packard-built, Merlin-engined machines which, known as BMk.10s, were ferried across the Atlantic for final fitting.

With a normal crew complement of seven, and powered by four 1,640 horsepower Rolls-Royce Merlin Mk.24 engines, the Lancaster cruised at some 210 mph (182 knots) – or at 140 mph (122 knots) if reduced to three engines. It operated at up to 22,000 feet, had a maximum take-off weight of 70,000 pounds and a range of 2,500 miles. As defensive armament it had eight 0.303 inch (7.7 mm) calibre machine guns, four in the tail, and two each in nose and dorsal turrets. The standard bomb load was 14,000 pounds or, having been modified, one 22,000 pounder bomb. For a comparison often made, the Flying Fortress's standard load was 6,000 pounds.

Bell 206B, Jet Ranger helicopter The 2004 publicity for the Bell 206B helicopter claimed, '*First in its class in safety, this aircraft ensures its crew will be back to fly another day.*' A 400 shaft-horse-power Allison 250-C20 series turboshaft engine typically gave a top speed of 132 mph (115 knots), a cruising speed of 115 mph (100 knots), an endurance of three hours, a range of 435 miles, and seating for between four and six passengers. The maximum external weight permitted was 1,500 pounds.

Blackburn Botha When the RAF first received the 1939 Botha – envisaged as a twin-engined reconnaissance torpedo bomber – several were lost in fatal accidents. The aircraft, normally carrying four crew, was then reallocated to training, lingering on as a target tug until 1944, by which time 580 had been built.

The Botha's twin 930 horsepower Bristol Perseus radial engines gave it a cruising speed of 212 mph (184 knots) with a ceiling of 17,500 feet. It had a maximum take-off weight of 18,450 pounds and was armed with a fixed 0.303-inch calibre machine gun firing forwards, and two others in a dorsal turret. In addition it could carry an internally stowed torpedo, or up to 2,000 pounds of depth charges or bombs.

Boeing B-29 Superfortress: F-13 variant The 1942 B-29 Superfortress development of the B-17 Flying Fortress was powered by four 2,200 horsepower Wright Cyclone Eighteen supercharged radial engines each driving a 16.5 foot, four-bladed, Hamilton propeller. This combination gave it a cruising speed of 220 mph (191 knots) and a ceiling of 33,600 feet. Its maximum bomb load was 20,000 pounds, but this reduced the range to 1,000 miles. At a maximum take-off weight of 133,500 pounds, and carrying a bomb load of 6,000 pounds, it had a range of 3,700 miles. The normal crew complement was ten, the bomber version being furnished with multiple 0.50-inch calibre machine guns and 20 mm calibre cannon. Futuristically, it was pressurised and had a computerised fire-control system!

The F-13 variant was a dedicated photographic machine in which specialists swelled the crew to 13. Although there were profound internal differences it was virtually identical in outward appearance to the B-29 bomber.

Boulton Paul Defiant The 1937 Boulton Paul Defiant, a low-wing, all-metal, twin-crewed fighter, had four 0.303 inch (7.7 mm) calibre machine guns mounted in a moveable dorsal turret. The weight and drag of the turret, however, prevented the aircraft from reaching an acceptable standard. Soon, too, German fighters realised that the Defiant had neither belly nor head-on protection, and in December 1940 it was withdrawn from daylight operations. Once it was re-employed as a night-fighter, however, and particularly when equipped with airborne-interception radar, it was to be rather more successful. Later it was to see further service as a target tug. Nearly 1,100 were built before construction ceased in 1943.

With a 1,280 horsepower Rolls-Royce Merlin Mk.20 engine it could cruise at 260 mph (226 knots) and climb to some 30,000 feet. It had a range of 465 miles and a maximum take-off weight of 8,424 pounds.

Bristol Blenheim In 1935 the machine ordered as a private runabout by air-minded Lord Rothermere proved to be faster than the RAF's latest fighters. By 1937 this had become the three-crewed, twin-engined Blenheim light bomber, of which 4,422 were eventually built, equipping 70 RAF squadrons. By 1939, however, the Blenheim had been far outclassed by the latest German types. Although quickly withdrawn from bombing operations the type continued to serve as a radar-equipped night-fighter, and later, as an advanced crew trainer.

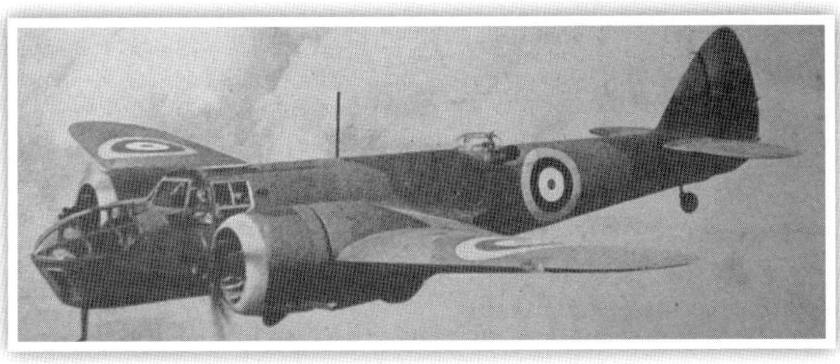

At a maximum weight of 14,400 pounds, and employing two 905 horsepower Bristol Mercury Fifteen radial engines, the Blenheim Mk.4 had a ceiling of 27,000 feet, a cruising speed of 198 mph (172 knots), and a range of 1,460 miles. Armed with two 0.303 inch (7.7 mm) calibre machine guns in a power-operated dorsal turret, with two remotely-controlled guns below the nose, and a fifth in the port wing, it could also carry a 1,300 pound load of bombs.

Cessna 150 The strut-braced, high-winged, tricycle-undercarriaged, two-seater Cessna 150 first flew in September 1957, after which over 30,000 were built with up to 20,000 still flying in 2010. Pleasant to handle, forgiving, viceless, and dependable, it is widely regarded as the most popular light aeroplane ever produced.

A 100 horsepower Continental 0-200A flat-four engine driving a two-bladed fixed-pitch propeller gave it an average cruising speed of 122 mph (106 knots) with 65-70 mph (56-61 knots) sufficing for virtually all manoeuvring. The ceiling was 15,000 feet, range 350 miles and the maximum take-off weight 1,500 pounds.

Consolidated-Vultee 32 Liberator The December 1939 long-range B-24 Liberator was immediately ordered by Britain and France, with Britain inheriting the whole order when France fell. The type was first used by the British Overseas Airways Corporation and Coastal Command. However, by September 1943 the Americans themselves had come to appreciate its value. Over 18,000 were built, with the production rate reaching one every 56 minutes.

Four 1,200 horsepower Pratt & Whitney Twin Wasp engines driving Curtiss three-bladed, electrically-driven, constant-speed propellers gave a cruising speed of 220 mph (191 knots) and a ceiling of 36,000 feet. The maximum take-off weight was 60,000 pounds, the range 2,500 miles, and the bomb load 8,000 pounds. Armament comprised 14 machine guns, four of 0.5 inch calibre in the nose, with two in a ventral turret, and four of the lighter 0.30 inch calibre in dorsal and tail turrets.

De Havilland Beaver The de Havilland of Canada L-20A Beaver first flew in August 1947, and being designed to operate from restricted spaces, had good short-field capabilities and a steep climb out. The Beaver could carry a pilot and seven passengers, had a cruising speed of 125 mph (109 knots), a ceiling of 18,000 feet, and a range of 676 miles. Some 1,700 were built and many were still being operated in 2010.

De Havilland DH9 During the First World War the de Havilland DH9 long-range day bomber of 1917 served with the Royal Flying Corps, and after the birth of the new Service on 1 April 1918, with the Royal Air Force. It also served with the effective precursor of Bomber Command, Trenchard's defiantly-styled Independent Force.

Despite a 240 horsepower engine the DH9 could only reach 10,000 feet with its full bomb load of 460 pounds and suffered accordingly from ground fire. It had a duration of four and a half hours, a maximum speed of 111 mph (96 knots), required just 112 yards to lift off, and at a landing speed of 57 mph (50 knots) a 160 yard run from touchdown. As defensive armament it had a forward-firing 0.303-inch calibre Vickers machine gun, and in the aft cockpit, one, or sometimes two, 0.303-inch calibre Lewis machine guns.

De Havilland DH9A The de Havilland 9A biplane bomber entered the RAF in June 1918 and was its standard day-bomber until 1931. Powered by either a 400 horsepower Liberty motor, or a 375 horsepower Rolls-Royce Eagle, it had a ceiling of 16,500 feet, and being operated in close formation, suffered only light losses in France. Nearly 900 were built with several hundred more following in the post-war years for service in Aden, Iraq, India, Egypt and Palestine. The DH9A also served many day-bomber units in the home establishment and from 1925 was widely used by the Auxiliary Air Force and many flying training schools.

Carrying 660 pounds of bombs, its endurance was just over five hours. When laden it cruised at 114 mph (99 knots) but its maximum speed was 126 mph (109 knots). As armament it had two 0.303-inch calibre machine guns; a forward-firing Vickers and a ring-mounted Lewis in the rear cockpit.

De Havilland DH82A Tiger Moth The 1934 improved Service version of the tandem two-seater biplane de Havilland Moth made an unassailable name for itself as a training machine at over 80 elementary flying training schools in the course of the Second World War.

Its 130 horsepower de Havilland Gipsy Major in-line engine gave the 1,770 pound machine a cruising speed of 93 mph (80 knots), a ceiling of 13,000 feet and a range of 300 miles. For solo flight the pilot sat in the rear seat to maintain the centre of gravity, while a hood facilitated dual instrument-flying training. The machine could be fitted with bomb racks, and indeed, saw operational service both as a communications aircraft before the fall of France, and as a maritime scout.

Although demanding to fly accurately, it had virtually no vices; just the same, it brooked no undue liberties. Examples of the type still flew in 2010, and it is almost universally spoken of reverentially. It might be held as sacrilegious, therefore, to recall that its cockpit was uncomfortable, and that it invariably gave a freezing-cold ride.

De Havilland DH89 Rapide De Havilland's light-transport biplane was developed from their other successful models during the 1930s, the RAF renaming it the Dominie and using it both in its design role as an eight- to ten-seater communications machine, and as a five- to six-seater navigation and radio trainer. Many civil Rapides were impressed into wartime service, notably with the Air Transport Auxiliary, the type subsequently being welcomed by many emergent and

re-emergent airlines in the early post-war years.

Driven by two 200 horsepower de Havilland Gipsy Queen in-line engines, the Rapide could cruise at 132 mph (115 knots) and attain 16,500 feet with a range of 570 miles. In all, 728 of the type had been built when production ended in mid-1946, and in 2010 some were still earning their living giving pleasure flights, notably at Duxford.

De Havilland DH90 Dragonfly In 1935 de Havilland brought out a five-seater, luxury-tourer version of the Dragon Rapide, building just 66. This machine, the Dragonfly, when impressed into RAF wartime service was employed in the same training and communication roles as the Rapide – the (original) RAF Dominie.

Powered by two 130 horsepower Gipsy Major engines, the Dragonfly cruised at 125 mph (109 knots) had a range of 624 miles, and a ceiling of 18,000 feet.

De Havilland Vampire The 1946 twin-boomed de Havilland Vampire, the third of Britain's jet aircraft, was a private de Havilland venture which had a lot in common with the Mosquito, the whole forward zone being predominantly aluminium-skinned wood. Nippy, yet essentially stable and easy to fly, the Vampire, and particularly the Mk.5 fighter-bomber variant (as at Strines, see above), also proved a useful stop-gap advanced trainer for Flying Training Command. A dedicated twin-seat, side-by-side trainer version, the T.11 ('Tee Eleven'), became available in 1950.

The T.11, with its Goblin 3 centrifugal-flow, turbo-jet engine developing 3,200 pounds of static thrust, had a ceiling of 40,000 feet, a maximum permitted speed of 523 mph (455 knots) and a medium-level cruising speed of 265 mph (230 knots). Its maximum take-off weight was 13,380 pounds, it had a range of 730 miles, and it was to remain in

RAF service until 1966. Although nominally a trainer it could mount two or four 20 mm calibre cannon, and provision was made to carry rocket projectiles or bombs.

Douglas C-47 Skytrain (Dakota) The 1935 Douglas DC-3, basically carrying 21 passengers and a crew of three, the doyen of air transports, was still flying commercially in 2010. Among its many variants the C-47 Skytrain, with a strengthened floor and wide cargo hatch, proved adaptable to a seemingly infinite number of tasks: not least transporting a jeep, as at Shelf Stones (see above).

The C-47's two 1,050 horsepower Pratt & Whitney Twin Wasp, air-cooled engines gave it a cruising speed of 207 mph (180 knots), a stalling speed of 67 mph (58 knots) and a ceiling of 23,200 feet. Maximum take-off weight was 25,200 pounds, and the range 2,125 miles.

Gloster Meteor Celebrated for being the only Allied jet aircraft to see service during the Second World War, the twin-jet Meteor first flew in March 1943. From July 1944, armed with four 20 mm calibre cannon, it was successfully deployed against the V1 Flying Bombs.

Early versions were powered by two Rolls-Royce Welland turbojet engines, each developing 1,700 pounds of static thrust to give a top speed of 415 mph (361 knots) and a ceiling of 40,000 feet. Later versions, using Rolls-Royce Derwent Eight engines, each giving 3,660 pounds of thrust, achieved nearly 600 mph (521 knots), with an initial climb rate of 7,350 feet a minute. Periodic updating included the provision of Martin Baker ejection seats. The type was phased out of service in the 1960s.

Handley Page Halifax Entering service in November 1940, the seven-crewed, twin-finned Halifax heavy bomber found favour with its versatility, for besides its design role it was employed in both the transport and maritime roles, also as an ambulance, a glider tug, and as a clandestine and paratroop-delivery vehicle.

A typical fit of four 1,615 horsepower Bristol Hercules Sixteen radial engines gave it a cruising speed of 215 mph (187 knots) and a ceiling of 24,000 feet. It had a range of 1,030 miles and a maximum take-off weight of 65,000 pounds. It could carry 13,000 pounds of bombs and mounted nine 0.303 inch (7.7 mm) calibre machine guns, one in the nose, and four each in dorsal and tail turrets.

An unfortunate characteristic of early Halifaxes was that fully-laden aircraft could enter an inverted, and effectively uncontrollable, spin. A retrospective

modification of the tailfin leading-edge shape from triangular to quadrilateral helped overcome this defect; 6,200 Halifaxes were built before its withdrawal from RAF service in 1947.

Handley Page Hampden The 1936 four-crewed Hampden, powered by two 1,000 horsepower, 9-cylinder, Bristol Pegasus Mark Eighteen radial engines, equipped ten RAF bomber squadrons at the outbreak of war. Here, though, is a case where propaganda-enhanced performance figures refuse to lie dormant. So, the Hampden's ceiling is frequently given as 19,000 feet, although Handley Page themselves only claimed 15,000 feet. At the same time, the company extolled their product's 'incredibly fast' 254 mph (221 knots) maximum speed. But although a 1942 source gives the cruise as 217 mph (189 knots), actual users found the workaday cruise to be nearer 130 mph (113 knots), with the least sanguine modern source proffering 167 mph (145 knots).

Irreconcilable figures aside, the Hampden's Handley Page leading-edge slots did give it a landing speed of just 73 mph (64 knots), and most sources agree that at a maximum take-off weight of 18,756 pounds it had a range of 1,885 miles with half a bomb load, reducing to 1,200 miles when the full 4,000 pounds was carried. As defensive armament it mounted two forward-firing 0.303 inch (7.7 mm) calibre machine guns, with additional twin mountings in both a dorsal and a rearward-facing belly position.

The Hampden showed up poorly against German fighters, however, and just a month into the war it was restricted to night operations, to leaflet dropping, and to minelaying. Although the Hampden was regarded as pleasant to handle, the crew found their positions cramped. Just the same, 1,432 were produced before the type was phased out in 1943.

Handley Page Heyford The 1933 twin-engined, basically four-crewed Heyford was a biplane-bomber of all-metal framed construction whose speedy 143 mph (124 knots) earned it the appellation, 'Express'. Indeed, unlikely as it seems, a No.102 Squadron Heyford was publicly looped during the 1935 Hendon Air Show! Interestingly too, it was held – admittedly by Handley Page – that, in comparison to a retractable undercarriage, the lighter weight of the streamlined but fixed undercarriage so minimised drag that it actually enhanced the Heyford's performance. Withdrawn from first-line service in 1939 the type still gave good value as a crew trainer until 1941, being stable, and pleasant to fly.

Powered by two 575 horsepower, Rolls-Royce Kestrel Mk.3 engines the Heyford had a ceiling of 21,000 feet and a maximum all-up weight of 16,900 pounds. Its full bomb load was 3,500 pounds, but with half that load it had an operational striking range of 920 miles. (Or as Handley Page 'spin' preferred, it 'carried a very large load of bombs for 2,000 miles'.) For defensive armament it carried three 0.303 inch (7.7 mm) calibre Lewis machine guns mounted respectively in dorsal, ventral, and nose positions. Production ended in July 1936, with 124 supplied.

Hawker Hart In air exercises the 1930 Hawker Hart bomber easily outstripped the latest fighters, so acting as a spur to the RAF's re-equipment. It also proved a highly versatile aircraft, seeing service with Auxiliary and Volunteer Reserve units. Typically powered by a 525 horsepower Rolls-Royce Kestrel engine it had a maximum speed of 184 mph (160 knots). It had a cruising speed of 145 mph (126 knots), a range of 430 miles, and a ceiling of 22,800 feet. Only in late 1939 did Harvards and Masters begin to replace the Hart, the consequence being that many distinguished Second World War pilots gained their wings on the type.

Hawker Hind By September 1938, as the RAF built up its strength, a total of 528 of the 1934 Hawker Hind day-bombers had been received. The was a refined version of the Hart, its fully-supercharged, 640-horsepower Rolls-Royce Kestrel engine giving it a maximum speed of 186 mph (162 knots). It had a ceiling of 26,400 feet, a range of 430 miles, could carry 500 pounds of bombs, and was armed with two 0.303-inch calibre machine guns; a Lewis to the rear and a forward-facing Vickers.

Hawker Hunter TMk.7 The Hunter, the first British-produced transonic fighter (see Glossary), was the RAF's mainstay first-line fighter from 1954, when it began replacing the Meteors and Sabres, until 1963, when it was replaced by the missile-armed, fully supersonic Lightning. Other air forces, however, carried on using it for many years, while not a few retired aircraft found eager private owners. Altogether, 2,000 were built.

Typically, the Hunter was powered by a Rolls-Royce Avon 203 turbojet developing 10,000 pounds of static thrust. This gave a maximum speed of 715 mph (621 knots)

at sea level, and Mach 0.95 at 36,000 feet. Fully laden, with drop tanks, it weighed 24,600 pounds. It had an initial climb rate of 17,200 feet a minute, took eight minutes to climb to 46,000 feet, and had a ceiling of 51,500 feet. It had a clean combat radius of some 320 miles, and a seemingly niggardly endurance of just 1 hour 18 minutes. Underwing drop tanks gave the Hunter a range of 1,840 miles, standard armament being four 30 mm Aden cannon together with underwing pylons for bombs or rockets.

Easy to fly – delightful, indeed – pilots who later transferred to Lightnings wistfully described the Hunter as 'the last fun fighter'.

Heinkel He111 The 1935 Heinkel He111 ('One-eleven') was blooded with the Condor Legion in the Spanish Civil War, and later in Poland, handsomely outstripping the opposing fighters. Over Britain, however, both its armament and performance proved inadequate, particularly as German fighters were unable to dwell long enough to provide meaningful support. From mid-September 1940, therefore, it was restricted to night-time operations.

Powered by two 1100 Junkers Jumo engines it had an average speed (collating various sources) of 250 mph (217 knots), a ceiling of 23,000 feet, and a range of 1,030 miles. Early versions had a crew of four, a bomb load of some 4,000 pounds and were armed with three 7.9 mm calibre machine guns mounted dorsally, in the nose, and in a belly turret. When production ceased in 1944 a total of 7,399 had been built.

Junkers Ju88 Fortunately for the Allies, German aircraft designers, like their British counterparts, frequently had changes forced upon them. The Junkers Ju88 bomber, for example, was envisaged as a fast, minimally-armed machine capable of targeting the whole of the British Isles. In the event, the Luftwaffe's insistence that it be used primarily as a dive bomber called for a more robust construction. This increased the weight and reduced the design speed and manoeuvrability, reductions which called for more defensive armament. The bitter pill for the German airmen being that the type was never actually used as a dive-bomber except when operating over water! The Luftwaffe received the Ju88 in 1939, 15,000 being built.

The type was adapted to many roles but representative were two 1,400 horsepower Junkers 211J liquid-cooled inverted V12 engines which gave a maximum speed of 295 mph (256 knots) and a ceiling of 26,900 feet. A typical maximum all-up weight was 30,865 pounds, with a crew of four comprising pilot, bomb aimer, top-gunner/radio-operator, and lower-gunner/flight-engineer. Machine-gun armament was one of 7.9 mm calibre under the control of the pilot, and three of 7.9 mm calibre and a 13 mm calibre fought by the gunners. Four 1,000 kilogram bombs could be carried.

Lockheed Hudson The twin-engined, twin-finned, five-crewed Lockheed Hudson was the military version of the Lockheed 14 Super-Electra airliner, the RAF ordering 200 in June 1938. The order caused outrage among those who believed that buying any but British aeroplanes was heinous, a pernicious lobby fortunately overruled on this occasion so that over 2,000 Hudsons were timely received, most flown over the Atlantic under their own power.

Intended as a navigational trainer, the Hudson was pressed into the maritime-reconnaissance and anti-submarine roles, carrying an air-droppable lifeboat while on air-sea rescue duties. It was also used as a bomber, and as a clandestine delivery vehicle for supplies and agents. Finally, once superseded as a first-line aircraft, it served as both trainer and transport.

Typically powered by two 1,100-horsepower Wright Cyclone engines, it cruised at 170 mph (148 knots), had an endurance of six hours, a range of 2,160 miles, and a ceiling of 22,000 feet. Carrying between 750 and 1,000 pounds of bombs, it mounted five 0.303-inch calibre machine guns; two below the nose, a moveable dorsal gun, and a pair in the rear turret. It could also carry two beam-mounted (waist) guns.

Miles Magister (RAF), Miles Hawk (Civil) The 1936 tandem-seated, low-winged, metal-skinned Miles Hawk monoplane so impressed the RAF that a Service version, the Magister elementary trainer, was ordered. A serious spinning problem was encountered, and solved, and other modifications were made. Basically, however, the Magister remained a Hawk in uniform. Certainly, after the war, many Hawks-become-Magisters were welcomed back by flying clubs and private owners.

Powered by a 130 horsepower de Havilland Gipsy Major One in-line engine, it had a cruising speed of 123 mph (107 knots) and a ceiling of 18,000 feet. It boasted wheel-brakes, power-operated flaps, and a tailwheel – as opposed to a skid – and could be flown solo from either seat, although the front seat was preferred. It had a maximum take-off weight of 1,900 pounds, a ceiling of 16,500 feet, and a range of 380 miles.

Unlike its contemporary, the Tiger Moth, it responded well in gusty conditions. On the other hand, unless controlled, the Hawk's wing would lift markedly in a crosswind. Then again there were trimming controls to master. Just two of the 'complications' that made it a good trainer. And there were no vices. Only, for all its good points, it never aroused anything like the affection engendered by the Tiger Moth.

Miles Master Mk.3 (see illustration on page 135) Showing a top speed of nearly 300 mph – just 20 miles an hour slower than the Hurricane – the 1935 private-venture Miles

Kestrel seemed the ideal machine for easing the transition from the Tiger Moth and Magister trainers to the first-line Hurricanes and Spitfires. However, following a pattern only too well established even by then, so many modifications were called for that in March 1939, when the emergent Miles Master trainer first flew, it was a full 100 miles an hour slower than the Hurricane. It did, however, retain handling characteristics similar to those of the new fighters.

After engine-fit problems, the Master Mark Three received the 825 horsepower Pratt & Whitney Wasp Junior radial, which gave the tandem-seated trainer a maximum speed of 232 mph (202 knots) and a cruising speed of 170 mph (148 knots) while retaining the 85 mph (74 knots) landing speed of earlier marks. It also had a ceiling of 25,000 feet, and at a maximum take-off weight of some 5,500 pounds, a range of 390 miles. Eventually 2,350 were built.

North American Harvard Some 17,000 of the 1935-vintage Harvards were built, far more than of any other wartime trainer. Both Britain and France ordered them, the French receiving over a hundred before being overrun, after which the Luftwaffe used them both for training and for familiarising pilots tasked with evaluating captured American machines. Though the Harvard remained in RAF service from January 1939 until 1955, the South African Air Force retained it until 1995!

A 550/600 horsepower, 9-cylinder Pratt & Whitney Wasp air-cooled engine driving a Hamilton two-bladed, two-position controllable-pitch propeller gave a cruising speed of 180 mph (156 knots). The landing speed was 63 mph (55 knots) with an initial climb rate of 1,350 feet a minute. The ceiling was 23,000 feet, the range 730 miles, and the maximum weight 2,260 pounds. Although popular, it was a demanding machine to operate – and therefore, a good advanced trainer – and required its pilot to be cognisant of both its handling qualities and its technical foibles.

North American P-51 Mustang Although the RAF took deliveries of the Mustang in 1942, the Americans themselves were slow to appreciate their home-constructed product. In consequence they lacked a long-range escort fighter when their European bomber offensive began later that year, losses soaring when their limited-range fighters had to turn for home. Only in December 1943, when the Rolls-Royce Merlin-engined,

Mustangs arrived in Europe, could the hard-pressed American bomber crews rely upon fighter support throughout an entire mission (see Glossary).

A typical version, the Mustang P-51B, was powered by a 1,520-horsepower Packard Rolls-Royce Merlin V-1650-3 liquid-cooled engine driving a three-bladed Curtiss electric constant-speed propeller. This combination gave it a maximum level-flight speed of over 400 mph (348 knots) and a ceiling of over 40,000 feet. Its range was over a thousand miles. As armament it mounted six or eight 0.5-inch calibre machine guns, or four 20 mm calibre cannon. It could also carry 1,000 pounds of bombs slung underwing.

North American Sabre (American, F-86) In the early 1950s the British-designed replacements for the by then outclassed Meteor and Vampire fighters were suffering many developmental problems, so the appearance early in the Korean War (1950-53) of the Soviet MiGs quite discomfited the Royal Air Force planners. Under a mutual defence agreement America made over 431 Sabre jets, many of which had been both developed and built in Canada.

The version the RAF received was powered by an Allison J47-GE-13 engine developing 5,200 pounds of static thrust which gave it a maximum speed of 679 mph (590 knots) and an initial climb rate of 7,250 feet a minute. Its all-up weight was 17,806 pounds. Most pilots found the Sabre a delight to fly and many expressed disappointment when it was replaced by the Supermarine Swift and by the early marks of the Hawker Hunter. Just the same, by mid-1956 Hunters had completely replaced the RAF's Sabres, both in Germany and in the UK.

Short Stirling The Short Stirling, the first of the RAF's heavy bombers, came into service in 1941. Shorts re-used their successful Sunderland-wing profile but Ministry requirements limited the span to 100 feet (not, as myth has it, to fit into RAF hangars, which were 125 feet wide). This, and similar modifications, detracted from the design performance to give the Stirling a ceiling of only 17,000 feet. It was, however, very manoeuvrable, and powered by four 1,650 horsepower Bristol Hercules Sixteen radial engines, had a maximum speed of 270 mph (235 knots), a cruise of 200 mph (174 knots) and a range – dependent upon bomb load – of up to 2,000 miles.

The Stirling carried 14,000 pounds of bombs, and had eight 0.303 inch (7.7 mm) calibre machine guns; four in a tail turret, and two each in nose and dorsal turrets. Its cockpit stood at a lofty 22 feet 9 inches above the tarmac.

It was popular with its seven- or eight-man crews but its bomb bay could not be adapted as bigger bombs were developed and it ceased bomber operations in September 1941. It was then very successfully employed in the glider-tug, clandestine-operations, and transport roles. In all, 1,759 bombers were produced, also 160 transport variants, production ceasing in November 1945.

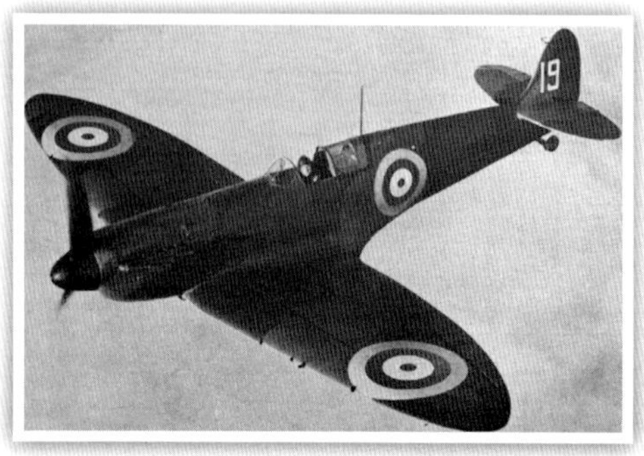

Supermarine Spitfire The Spitfire first flew in 5 March 1936 and by October 1947, when production ceased, had metamorphosed through over a score of variant Marks with some 23,000 built. The early Spitfire, like that at Hooton Roberts, was powered by a 1,030 horsepower Rolls-Royce Merlin Mark Two in-line engine, driving a wooden two-bladed, fixed-pitch propeller. This gave it a maximum level speed of 355 mph (308 knots) and a cruising speed of 265 mph (230 knots). It took 6.2 minutes to reach 15,000 feet and its ceiling was 34,000 feet. The undercarriage and flaps had to be manually operated and due to supply difficulties only four – rather than the planned eight – 0.303 inch (7.7 mm) calibre machine guns were installed. None of this would give a complete picture, however, unless the superb handling qualities of the machine were mentioned.

V1 (*Vergeltungswaffe*) Flying Bomb The German *Fern Ziel Geraet* (Long-Range Target Apparatus), their *Vergeltungswaffe 1* (Reprisal Weapon Number One), was a normally ramp-launched pilotless flying bomb which cruised at 410 mph (356 knots), had a ceiling of 10,000 feet, and a range of some 150 miles. It was powered by a singularly-sounding pulse-jet engine which, at a preset range, cut out, the cessation of the sound indicating to the initiated below that an explosion was imminent. The range was vastly extended when V1s were mounted below the wing root of Heinkel He111 bombers and air launched. It is said that of 10,000 V1s launched against England, 7,000 landed on the mainland. For newer generations it might be politic to reiterate that while the V2 was a rocket, the V1 (these days, too-often termed 'V1 rocket') was a pilotless pulse-jet aircraft.

Vickers Armstrong Wellington In designing the 1937 Wellington, the celebrated Barnes Wallis used repeated junctions of Meccano-style alloy channels to form a cocoon of great strength. This 'geodetic' structure (additionally see Glossary) was

then covered with doped fabric.

A typical power fit was two 1,500 horsepower Bristol Hercules Eleven radial engines which gave a ceiling of 19,000 feet and a maximum speed of 235 mph (204 knots). Representative cruising speeds vary with source, ranging from 232 mph (202 knots) to 166 mph (144 knots), but Brooklands, the Vickers' Museum, suggest 173 mph (150 knots) with a normal bombing altitude of 12,000 feet. The bomb load was 4,500 pounds and the armament eight 0.303 inch (7.7 mm) calibre machine guns; four in the tail turret, two in the beam, and two in the nose. The German defences, however, soon took the Wellington's measure, after which it was switched to night bombing and maritime operations.

The operational crew of four comprised a pilot, a navigator/bomb-aimer, a wireless operator/air gunner, and a rear gunner. A total of 11,461 Wellingtons were built, the type continuing in service until 1953 using aircraft dedicated to both the pilot and navigator training roles.

Zenair Zodiac The Zenair Zodiac self-build kit aircraft of 1984 was designed as an all-metal, side-by-side two-seat, fixed tricycle-undercarriaged machine, the recommended power plant being the Lycoming 0-235 piston engine. The specification had the Zodiac carrying a pilot and passenger to an all-up weight of 1,220 pounds (600 kg) with a never-exceed speed of 161 mph (140 knots), a cruise of 130 mph (111 knots) and a basic stalling speed of 44 mph (37 knots). The range quoted was 715 miles (1,150 km) with a ceiling of 12,000 feet. Its history records rather a large number of wing failures.

Glossary

Aviation

Air Plot: essentially, a navigational method which records the aircraft's progress using a line (a 'vector') drawn to scale to represent the machine's airspeed and direction. The best-known wind is then applied as a second vector to produce the present position. The procedure is updated by visual sightings or radio-bearings.

Bar (to a gallantry award): this term indicates a further award of a gallantry decoration and is represented by a rosette sewn to the ribbon. Not to be confused with the 'medal bar' or 'medal clasp' (denoting campaigns) running across the ribbon of a general service medal.

Circuits and Rollers: a 'circuit' – circuits and landings – involves taking off into wind, turning downwind parallel to the runway, flying past the airfield, then turning back, touching down, and rolling to a stop before clearing the runway. A 'roller' (circuits and bumps), on the other hand, requires the pilot to touch down but, without coming to a halt, to put on full power, reconfigure the aircraft for flight, and take off again.

Darky: an emergency homing system making a benefit of the very short range of wartime voice-radio sets. Merely hearing an aircraft call meant that it was quite close to the listening station. Telling the aircraft where the listening station was located, therefore, might well give the crew information enough to re-start their own navigation. Conversely, the aircraft could be directed towards the nearest airfield, each telephone-alerted listening station en route refining the direction to fly.

Dorsal (turret): a turret mounted on the top – the back – of the aircraft.

Enthusiast-compiled lists: these were very valuable in research. Entries which proved spurious are acknowledged to save other walkers from fruitless searches.

Forced-landing/precautionary landing: a forced-landing is a set-down caused by a malfunction which gives the pilot no option but to alight. A precautionary landing is one where the pilot decides that it is politic to put down, so permitting the choice of a suitable site.

'g': acceleration due to gravity. Any high acceleration manoeuvre – change of direction, effectively – results in a change of weight, or of centrifugal force, which is categorised as measuring so many 'g'.

Gee (Ground Electronic Equipment): an airborne radar system. To obtain a fix, radar-derived signals were plotted on a lattice chart. For use as a bad-weather let-down aid, heights were calculated against ranges to give a glide path.

Geodetic: the structure developed by aircraft designer Sir Barnes Wallis and employed in the Wellington bomber. Essentially, it comprised triangular grids made up of aluminium strips to form a mutually-supporting shell of great strength. More properly, the component parts formed 'geodetic' curves (parts of a circle) on the structure, each element taking the shortest line across the curved surface.

Hairpin: a reversing-direction manoeuvre in which the aircraft climbs, turns one way long enough to widen the angle, then turns steeply back the other way until it can dive along the reversed line of flight.

Mach number: named after the Austrian physicist Ernst Mach (1838-1916), this refers to the speed of an aircraft in relation to the speed of sound. So, an aircraft moving at twice the speed of sound travels at Mach 2, one at just 0.95 of sonic speed, Mach 0.95.

Ministry of Defence: created in 1971. Formerly its responsibilities in the matter of crash sites lay with the Air Ministry and the Ministry of Aircraft Production (later, Supply). The existence of the Protection of Military Remains Act 1986, order 2008, has to be acknowledged, but though forbidding unauthorised tampering with crash-sites, it says nothing of MOD having any higher responsibility to the countryside.

Mission (terminology)**:** throughout the era embracing the Second World War, offensive flights against the enemy were termed missions by the United States Army Air Force and operational sorties – or Ops – by their RAF counterparts. The standard operational tour required from RAF crews may be taken as thirty.

Pilotage: navigating by mapreading. Although seemingly dated, the word was to be employed by the up-to-the-mark crew of the Apollo 8 moonshot in 1968, who recorded having had, 'difficulty in "pilotage", that is, in trying to plot our path on the map of the back side of the moon'.

Radar Service advisory/control: essentially, under radar advisory service an aircraft is guided, but arranges its own terrain clearance. Under radar control, the radar station assumes full responsibility for the navigation.

Special Operations Executive (SOE): a volunteer organisation set up in July 1940 to carry out sabotage and subversion behind enemy lines. Churchill described its purpose as being to 'set Europe ablaze'.

Standard Beam Approach (SBA): in essence, this was a radar landing aid which transmitted a 30 mile long, very narrow radio beam down the extended centreline of the runway. This told a pilot receiving the aural 'on-the-beam' signal that he was somewhere along the projected centre line of the runway. To furnish an exact location *along* the beam, an 'Outer Marker' radio beacon was sited at a known distance from touchdown. This sent a coded signal vertically upwards to tell an inbound pilot that he should commence his final approach, descending at a rate of 600 feet a minute.

Stick (control column): certainly, from the fifties this was always the preferred term among pilots; 'pole' was equally acceptable but somewhat informal, 'joystick' almost antediluvially archaic, and 'control column' too pedantic even for Central Flying School. So stick it is, even where the aircraft in question had a wheel, or a yoke.

Transonic: relating to speeds close to or equalling the speed of sound.

Very pistol, sometimes Very's pistol: a breach-loading, wide-bored signalling handgun firing cartridge flares of various colours, named for its 1877 inventor, American naval officer Edward Wilson Very.

Wreck: a misnomer employed by air-crash enthusiasts seeking an elegant variation on 'air crash'. Any class of aircraft may be wrecked if it is on the ground, but the nautical model, though legitimately transferred in the railway context, is misemployed for machines which come to grief in flight. Notwithstanding this, mainstream aviation usage embraces wreckage as a synonym for debris.

Walking

Abeam: lying at right angles to the line of march. All things being equal, if an established path is followed until a site is directly off one's shoulder, then the least amount of rough walking is required to reach that site.

Col: a depression or lowered section on a hill range.

Clough: a water-carved ravine leading from an upland peat moor. Most Peakland cloughs leave the rim in a steep river of boulders which look daunting but offer many routes through. Any too-steep pitches can be circumvented by backtracking a few yards.

Convex: used of a slope. From the rim the slope bellies outwards, so preventing a view of the ground immediately below.

Degrees magnetic: measuring a track on the map will give the true direction. Adding five degrees will give the direction to set on the compass. So, a measured track of 070° is set on the compass ring as 075°. (Purists – and enthusiasts – will blanch, but the rest of us will be tramping heather for no more than half a mile or so.)

Grough: a water-carved gully in an upland peat moor, often 20 feet deep. The peat is soot-black, and just as greasy. Egress, however, is always to be found within a few yards. Groughs, when going in the right direction, can afford easy passage.

Hag: the basically firm heather or bilberry stretches of ground left by the deep-cutting groughs.

Yards/Metres: again, let purists go pale, but to the workaday walker these are interchangeable up to half a mile or so.

Acknowledgments

To the pioneering, joint authors of the two *Dark Peak Aircraft Wreck* books (1979 and 1982) who paved the way for all walkers puzzled by metal fragments chanced upon while traversing the Peakland Moors. To Ron Collier, 1935-2010. One of his field companions remembers how he tramped the moors in the seventies 'with nothing but a compass, hearsay, and myths to go on, so that locating a wreck often took him weeks'; additionally Ron devoted 25 years to the Air Training Corps and qualified as a private pilot; also to Susan, his wife, who bore the burden of his so-protracted, failing illness. To Roni Wilkinson, who, as an author of boys' stories, set the tone for Ron's findings, serialising the material in the *Barnsley Chronicle* and subsequently joining Pen & Sword Books Ltd.

To veteran air-crash researchers John Ownsworth and Alan Jones (a noted aviation artist), both of whom furnished much extra-archival detail. This also applies to David W. Earl, author of the *Hell on High Ground* books.

To Malcolm Barrass, whose superlative website *Air of Authority* (www.rafweb.org) is a never-failing and utterly dependable source.

To Mr Ian Burgess, of Bury, and Séan Moran, of Wirksworth, who supplied 'links' enabling the quality of enthusiast web-forum observations to be assessed.

To Mr Brian Thompson, who re-evaluated two sites on Howden Moor, and to Mr Geoff Eyre, National Trust tenant of Howden Moor, for associated research.

To the several hundred folk interviewed, particularly from busy farming families, who gave their time to the research for this series.

To the RAF Museum, the Imperial War Museum, and to the British Library, for assistance with transcribing wartime map references to modern co-ordinates.

To the photographic staff at ASDA, Spondon, who, if irreverent, gave unstinting assistance.

To the traced copyright holders authorising the use of their photographs: Mr Gordon Miller, former Area Ranger for Kinder (front cover inset); Nicola Hunt, intellectual property rights copyright unit, MOD; archives staff, Imperial War Museum; Judy Nokes, licensing adviser, HMSO (Crown Copyright/MOD); John Ownsworth, for photographs used by Ron Collier; archives staff, Royal Air Force Museum; Mike Stowe, American crash reports; Graham Hutchinson, The Stirling Trust; Julian Temple, archivist, Vickers' Brooklands Museum, Weybridge; Roni Wilkinson, Pen & Sword Publishing. Craving the indulgence of those for whom all contact attempts have failed.

Despite such inestimable assistance, any errors remaining, and all opinions expressed, are my own.

Pat Cunningham DFM

Selective Bibliography

Air Ministry (1937) *Royal Air Force Pocket Book, AP1081*. London: HMSO
Air Ministry (1941) *Air Navigation Volume 1, AP1234.* London: HMSO
Air Ministry (1943) *Elementary Flying Training, AP1979A.* London: HMSO
Air Ministry (1948) *The Rise and Fall of the German Air Force (1931 to 1945).* London: HMSO
Air Ministry (1954) *Flying, Volumes 1 and 2, AP129.* (Sixth edition). London: HMSO
Air Ministry (1960) *Flying Instructor's Handbook,* AP3225D. London: HMSO
Air Ministry (1960) *Pilot's Notes Vampire T.11*. London: HMSO
Barrass, Malcolm (2005) *Air of Authority* (www.rafweb.org), (RAF organisation)
Bennett, D.C.T. (1936) *The Complete Air Navigator.* London: Pitman
Buxton-Doyle, Margaret (2003) *The Longdendale Valley.* Stroud: Tempus Publications
Collier, Ron; Wilkinson, Roni. (1979, 1982) *Dark Peak Aircraft Wrecks 1 & 2.* Barnsley: Pen & Sword
Cunningham, Pat (2005-6) Peakland Aircrashes Series: *The South* (2005); *The Central Area* (2006); *The North* (2006). Ashbourne: Landmark Publishing
Director of Flying Training RAF (1955-1970 various) *Air Clues.* London: MOD
Fellowes, P. F. M. (1942) *Britain's Wonderful Air Force.* London: Odhams
Hammerton, J. (1943) *ABC of the RAF.* London: Amalgamated Press
Handley Page Ltd (1949) *Forty Years On*. London: Handley Page
HMSO (1942-1943) *Aircraft Recognition.* London: Sampson Clark
Hurst, Ian; Bennett, Roger (2007) *Mountain Rescue*. Stroud: Tempus Books
Lamplugh, A.G. (1931) *Accidents in Civil Aviation.* Royal Aeronautical Paper, Institution of Aeronautical Engineers, 29 October 1931, London
Monday, David (1982) *British Aircraft of World War II*. Chancellor Press: London
Office of Public Sector Information (OPSI) (2008) *Protection of Military Remains Act 1986, order 2008*. London
Phelps, Anthony (1944) *I couldn't care less.* (Air Transport Auxiliary) Leicester: Harborough
Saville-Sneath, R.A. (1945) *Aircraft of the United States, Volume One*. London: Penguin
Smith, Peter J.C. (1988) *Flying Bombs over the Pennines.* Manchester: Neil Richardson
Stewart, Oliver (1941) *The Royal Air Force in Pictures.* London: Country Life
Sturtivant, Ray; Page, Gordon (1999) *'Air Britain Listings' series.* Old Woking: Unwin
Thetford, Owen (1958) *Aircraft of the Royal Air Force 1918-58.* London: Putnam

Index

A

B

C

E

Index

Index